THE BEE GARDEN

THE BEE GARDEN

HOW TO CREATE OR ADAPT
A GARDEN TO ATTRACT AND NURTURE BEES

Maureen Little

SPRING HILL

Published by Spring Hill, an imprint of How To Books Ltd.
Spring Hill House, Spring Hill Road
Begbroke, Oxford OX5 1RX
United Kingdom
Tel: (01865) 375794
Fax: (01865) 379162
info@howtobooks.co.uk
www.howtobooks.co.uk

First published 2011
Reprinted 2013

How To Books greatly reduce the carbon footprint of their books
by sourcing their typesetting and printing in the UK.

British Library Cataloguing in Publication Data
A catalogue record of this book is available from the British Library.

ISBN: 978 1 905862 59 7

Produced for How To Books by Deer Park Productions, Tavistock, Devon
Designed and typeset by Mousemat Design Ltd
Printed and bound by in Great Britain by Bell & Bain Ltd, Glasgow

NOTE: The material contained in this book is set out in good faith for general guidance and
no liability can be accepted for loss or expense incurred as a result of relying in particular circumstances on statements made in the book. Laws and regulations are complex and liable to
change, and readers should check the current position with relevant authorities before making
personal arrangements.

Contents

Introduction   xi

Acknowledgements   xv

1 Bees and Their Needs   1

Honeybees   1

Different types of honeybee   8

The colony   9

Pollination   11

Moving bees   13

What do bees need?   14

Bees need plants   14

Bees need pollen   14

Bees need nectar   20

Bees need water   26

Bees' vision   28

The pollen basket   30

The proboscis   31

The waggle dance   32

How do plants attract bees?   34

What plants are best for bees?   35

2 Creating and Maintaining a Bee-friendly Garden   43

Organic gardening   44

Ground preparation   49

Conditions in your garden   50

Growing in containers   52

Planting trees and shrubs   53

Watering   54

Tidying up the garden   54

Mulching   55

Weeding   55

Pest control   55

Increasing plant numbers   55

Cuttings   56

Division   60

Sowing seeds   60

Labelling   62

Urban gardening   63

Gardening in the country   65

Gentle chaos   66

3 Plants and Gardening Jobs Season by Season   69

Spring – March to May   69

Spring plants for bees   70

Spring jobs for bee-friendly gardeners   78

Summer – June to August   84

Summer plants for bees   86

Summer jobs for bee-friendly gardeners   102

Autumn – September to November   106

Autumn plants for bees   107

Autumn jobs for bee-friendly gardeners   113

Contents

Winter – December to February   116

Winter plants for bees   117

Winter jobs for bee-friendly gardeners   117

4 **Other Flying Insects**   **121**

Bumblebees   121

Most common bumblebees   122

Solitary bees   126

Hoverflies   127

Wasps   127

Plants for bumblebees   130

Homes for bees   130

5 **Gazetteer of Plants Attractive to Bees**   **137**

Plant families   137

Annuals and biennials   141

Flowering times of selected annual and biennial flowers   149

Herbaceous perennials and bulbs   150

Flowering times of selected perennial flowers and bulbs   172

Shrubs and climbers   173

Flowering times of selected shrubs and climbers   183

Trees   183

Flowering times of selected trees   191

Edible fruits   191

Flowering times of selected edible fruits   201

Herbs   201

Flowering times of selected herbs   209

Wild flowers   210

Flowering times of selected wild flowers   217

6 Planting Plans   219

Culinary herb garden	219
Planting plan for herb garden	220
Potager	222
Planting plan for large potager	228
Planting plan for small potager	230
3 seasons border	232
Planting plan for 3 seasons border	235
Flowering times of plants in 3 seasons border	236
Wild flower garden	237
Planting plan for wild flower garden	239
Flowering times of flowers in wild flower garden	240

Appendices

1. Annuals and Biennials of Value to Bees	241
2. Perennials and Bulbs of Value to Bees	242
3. Shrubs and Climbers of Value to Bees	244
4. Trees of Value to Bees	245
5. Edible Plants of Value to Bees	246
6. Herbs of Value to Bees	248
7. Wild Flowers of Value to Bees	249

Further Reading	250
Useful Addresses and Websites	252
Glossary	254
Index of Common Names of Plants	258
Plant Index	263
Index	268

Introduction

... let gardens grow, where beelines end,
sighing in roses, saffron blooms, buddleia;
where bees pray on their knees ...
Carol Ann Duffy, 'Virgil's Bees'

As gardeners, we cannot help but be aware of the environment and the responsibility we have for the creatures that live in it. There has been a huge increase in the interest shown in the difficulties facing the honeybee, in particular. Many people, even if they do not wish to keep bees themselves, are asking what can be done on an individual basis to help the bee. This book is a response to that request. I hope that by reading this book we can all be inspired to play our part by providing a bee-friendly environment – no matter how much gardening space and/or time we have – where bees will come for sustenance, as well as (as Carol Ann Duffy submits) to pray.

One reason for looking at honeybees in particular is that, generally speaking, what is suitable in the way of plants and flowers for honeybees is bound to be acceptable to other flying insects, especially bumblebees and solitary bees.

As well as being a gardener, I also help my friend and bee mentor, Toady, look after his bees. I am not really sure if I am a gardening bee-keeper or a bee-keeping gardener (it depends what I am doing at the time, I think). Either way, both activities have a place in my heart for two reasons. First, they both provide me with food for my body in the shape of fruit and vegetables, and honey. Second, and more importantly, when I am tending my garden or the bees, they provide food for my soul and allow me instances of utter contentment that no material goods could ever afford.

This was brought home to me on one occasion when I was helping Toady check a number of hives which were on a rape crop. It was a beautiful summer's day and the bees were contented, buzzing around collecting nectar and pollen and not at all disturbed by my presence. I paused before moving on to the next hive, taking a deep breath and savouring the moment. Then another sound layered itself above the buzzing of the bees. A skylark had begun its heart-lifting song, reaching ever higher. All else melted away: it was just me, the bees, the skylark – and a privileged glimpse of heaven on earth. I know that both gardeners and bee-keepers alike will be nodding in acknowledgement.

The majority of people who keep bees today do so as a hobby, echoing, if not directly reflecting, the time when country folk had one or two hives in their gardens, hens in the run, and a pig in the sty. (Back then, of course, keeping livestock was part of the family economy: no animals meant no meat on the table.) My friend, Toady, however, keeps bees on a commercial basis, but the principles of bee-keeping are the same whether you have one hive, a hundred hives, or a thousand hives. They are living creatures and must be nurtured and cared for. Above all you must make sure they can access sufficient food and water to sustain themselves and the baby bees – the brood. It is true that commercial bee-keepers will seek out agricultural and horticultural crops for their bees simply because of the numbers involved, but those same bees will also forage in gardens if they are close enough to the hive and have sufficient food growing in them to make their journey worthwhile. This is where the gardener can play a part. By knowing what bees require and then growing plants which supply those needs, the gardener can perform a vital role in the survival of the honeybee, and other bees.

In Chapter 1 we shall be looking at why honeybees are so important and what makes them 'stand out from the crowd'. We shall also consider what they need to thrive and how they detect and access those requirements. Finally we shall touch on what sorts of plants are best suited to provide those needs.

In Chapter 2, I will cover the sort of 'universal' gardening jobs that can be done, especially if you want to create and maintain a bee-friendly garden. The gardening jobs are generally suggestions with a little detail, but if you would like more specific advice, the Royal Horticultural Society website (the address of which can be found in the Further Reading section at the end of the book) is one of the best starting points.

In Chapter 3 there is a season-by-season account of what bee-friendly plants are in flower and what jobs the gardener can be doing during these times.

In case you think I have forgotten about some of the other buzzy insects like bumblebees, you will find some information about these in Chapter 4.

Chapter 5 consists of a brief overview of plant families and looks, too, at which ones are particularly rich in bee-friendly plants. Following this there is a detailed gazetteer of selected bee-friendly plants, arranged by type of plant, in seasonal subsections. For each type of plant you will also find a chart which illustrates in graphic form the flowering times. This way I hope it will be easier for those of you who want to plan a planting area from scratch to choose different sorts of plants which flower at varying times so that food can be provided for the bees for as long a period as possible.

Finally, Chapter 6 contains five illustrative planting plans – a culinary herb garden, a potager (large and small), a '3 seasons' border and a wild flower garden – to give you some concrete ideas of how to put it all into practice!

Hardiness Zone

My experience as a gardener is restricted to the British Isles so all the recommendations I make and examples I give in this book are based on this. Our climate has been categorized as falling generally within Hardiness Zones 8a or 8b so if you are gardening outside the British Isles, adjustments must be made.

Latin names

In nearly all cases I have given the Latin name of the plant first with the common name in parentheses. The reason for doing this is that common names for plants can vary from region to region (a bluebell in Scotland is not the same as a bluebell in England, for example), so knowing the proper, undisputed Latin name is invaluable, especially when it comes to looking for plants in a nursery or online. The exceptions are names for vegetables: there is an unwritten convention that common names are used rather than the Latin ones. You will find a list of common names of plants and their Latin equivalents towards the end of the book.

Acknowledgements

For Georg, who helps me tend my garden,
and Toady, who taught me about bees.

If the honey that the bees gather out of so many flowers of herbs, shrubs and trees …
may justly be called the bees' honey … so may I call it, that I have learned and
gathered of many good authors (not without great labour and pain), my book.

William Turner, 1562

I do claim this book as mine. However, as has been rightly pointed out before, the *writing* of a book may be a solitary venture but the *creation* of it is a collaboration. My fellow collaborators, and others from whom I have gained inspiration, are many and talented (any mistakes that you may find are entirely mine):

My friend and bee mentor, Toady, without whom I would never have been introduced to the pure joy of bee-keeping and would still be trying to find what I have spent the greater part of my life looking for;

The 'alchemists' at Spring Hill who have turned my base materials into a precious book (especially Giles Lewis and Nikki Read);

All my fellow gardeners and bee-keepers, who care about bees as much as I do (in particular Paula Clatworthy, who was the first to reserve a copy of this book);

And, most importantly, my family, especially Georg, Becca and James (never forgetting Susie). They are all such a part of my life that without them my garden would be a wasteland and the honey would not taste as sweet.

In addition I would like to say thank you to:

My friend, Lancastrian and plantswoman Tricia Brown at 'Perennials by Design', who lent me her flowers!

Ann Turner, Principal of Myerscough College, Lancashire, who allowed me to take photographs of the college grounds, some of which are included in this book.

1
Bees and Their Needs

Everyone knows what a bee looks like. It is a round, fluffy, ball-shaped insect with bands of black and yellow that theoretically is not supposed to be able to fly – isn't it? Yes and no. That description most closely fits the largest bumblebee found in Britain – *Bombus terrestris* or the buff-tailed bumblebee. There are in fact some 250 different species of bee in Britain, including 24 bumblebees, various honeybees and solitary bees. Add to that other flying insects such as hoverflies and wasps and you have a vast number to get confused about!

Honeybees

The focus of this book, however, is on honeybees – the others have not been forgotten and we shall be looking at bumblebees and other flying insects that we are likely to come across in our garden in a later chapter. All have a part to play in the maintenance of ecological balance (although some people, me included, would say that we can manage quite nicely without the wasp, thank you very much). But it is the honeybee that takes centre stage in this book. Why, then, should we pay particular attention to the honeybee? What makes it so special that I have singled it out from bumblebees, say? What do honeybees do that the others do not? The main, and vital, difference is that honeybees can be managed by man. Millennia ago, humans found honey to be delicious and have been gathering honey and farming bees ever since. (Honeybees are not the only insect that man has made use of, of course; think of the silkworm.)

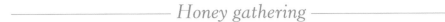

Honey gathering

The earliest record of honey gathering, rather than bee-keeping, can be found depicted in Middle Stone Age rock art in the 'Cave of the Spider' near Valencia in Spain. Honey gathering still takes place twice a year in the foothills of the Himalayas in Nepal. Men gather chunks of honeycomb from the wild bee colonies while clinging precariously to rope ladders let down from the top of the cliff – not a task for the faint-hearted, or if you are scared of heights. It was in 1978, however, that Dr Eva Crane, the renowned bee-keeper and physicist, unearthed beehives made from vegetation that had been coated in mud or clay in the Nile valley. They were identical to the hives represented in stone bas-relief found in the Temple of the Sun in Lower Egypt which were dated to 2400BC. Bee-keepers, then as now, were trying to manipulate the bees, persuading them to make as much honey as possible without compromising the welfare of the colony. After all, bees do not make honey for the benefit of humans. Honey is bee food.

Farming bees

When man initially started keeping bees he imitated the bees' native habitat by using tree trunks, willow or straw skeps and even clay pots. In these types of situations, as in the wild, the bees will build honeycomb vertically, starting at the top and gradually working their way down. The resulting comb is often elliptical in shape and there will be several built side by side. As an experiment, we housed a swarm that we had captured in a hive without the usual set of frames on which the bees would normally build up their comb. After a few days we took the lid off the hive and discovered that the bees had started building a comb as they would have done in the wild (Figure 1). (In case you're concerned about the fate of the colony, we provided new frames before the queen started laying in the newly created comb and shook off all the bees into their 'new' home where they have built up very nicely. We removed the comb from the lid and have kept it for teaching purposes.)

Skeps

In Britain bees would have been kept in skeps (Figure 2). These were rounded, conical containers made of coiled straw which were much smaller than modern beehives. If the bee-keeper wanted to increase the size of the skep then an additional ring, or eke, of coiled straw could be attached to the bottom. (Hence the term 'to eke out'.) Most 'everyday' bee-keepers with only a few skeps would keep them under a wooden lean-to at the side of the cottage. Larger houses would have specially designed structures to house the skeps. These usually took the form of a wall with a number of niches in it, open only at the front, each of which was large enough to house one skep. These 'bee boles', as they are called, were sometimes quite elaborate affairs, designed more for aesthetic than utilitarian reasons.

The major drawback of keeping bees in skeps was that the entire colony was forfeited to access the honey. The cleaned-out skeps would be replenished each year with swarms from a 'parent' hive which would be maintained solely for that purpose and not to take honey from. The introduction of the modern beehive in the 1800s changed all that. A system of removable frames was established which meant that honey could be harvested without sacrificing the colony. The history of the beehive and bee-keeping itself are fascinating topics but beyond the scope of this book. Have a look in the Further Reading section for some suggestions on books to look at.

Figure 1 Naturally formed honeycomb

Feral bees

You may think that unless there is a beehive nearby you are not likely to encounter honeybees in your garden. True, you are more likely to come across them if there is a hive in the vicinity, but a feral, or wild, honeybee colony may have established itself in your locality, especially if you live near a park or woodland. In the wild, honeybees nest in cavities in hollow trees, buildings, rocks or caves.

During a recent visit to a garden in Cumbria I saw a large number of honeybees, busily working some *Nepeta* (catmint) (page 7) and *Hydrangea aspera* 'Macrophylla' (page 9). My first thought was that there were some beehives nestling in a corner of the garden, out of sight of the public, but when I found the flightline of the bees it led towards a house and disappeared around the corner. This called for some investigation. I did not don my deerstalker and pull a magnifying glass out of my pocket but I did follow the bees, taking care not to tread on the beautifully tended flower beds, nor to venture anywhere marked 'Private'. When I rounded the corner of the house I saw the bees disappearing into the ventilation grille at the base of a chimney pot in a disused building, some seven or eight metres up. These were obviously feral bees that had found a home very much to their liking. I wondered how big the nest was, and how much honey they had made. But another point of interest is that the bees that I had observed were very clean, and free from varroa mite (a parasitic mite that can weaken, decimate and ultimately annihilate a colony). The absence of varroa in feral colonies is unusual. There is circumstantial evidence that the carbon and sulphur present in soot has been known to reduce the instances of varroa, but it is unproven – perhaps, however, the cleanliness of bees that I came across is a further proof of this claim? But I digress.

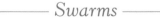

Swarms

If feral honeybees do visit your garden then the chances are that they are from a managed colony that has swarmed. This is where, for various reasons – the main one being lack of space – a new queen has emerged in an existing hive and the old queen has moved on, taking a large retinue with her. (It can be argued that if the bee-keeper is managing the hive well then swarming can be prevented, but even the most attentive of keepers can overlook the presence of a new queen cell in a hive, and the inevitable will happen.) If the bee-keeper is lucky s/he may be able to catch the swarm but

Figure 2 Straw skep

mostly it will evade even the most determined of chasers! Here, in the words of Toady, is a (somewhat exaggerated, in my view!) description of me catching my first swarm:

> This year Maureen caught her first swarm. It was a picture of delight: there she was, surrounded by about 20,000 bees (give or take a few!) flying mainly above her head, shouting at the top of her voice as to where they were going. She bolted off like a spring lamb chasing her swarm. I settled down in the corner of the field with a flask of tea, a slice of cake and the joy of watching Maureen bouncing from field to field, over hedgerow and stream, until this triumphant voice shouted, 'I've got them!' I had spent the good part of thirty years losing swarming bees, whilst Maureen caught her first swarm straight away, boxed them into a hive and proceeded to gloat for the rest of the week. Less said the better.

Swarms do not always choose a convenient place to settle and you can come across one in the most unlikely of places. On one occasion Toady and I had business in our local city. While he was in the bank I made my way to the chemist. Passing by a large container planted up with shrubs and flowers in the main shopping street, I spied a number of bees buzzing around, just above shoppers' heads. Then I saw it. This was not just a swarm; this was a well-known high street clothing and food shop swarm. It had landed on the main stem of a topiaried conifer in the planter just outside that particular store. On a visit to the local bank and chemist, even the most diligent of bee-keepers does not normally go equipped for catching a swarm. After fetching Toady, I got hold of a cardboard box from a bemused assistant in the store and watched as Toady, unsuited, and with bare hands, gently lifted the swarm from the shrub and lowered it into the box. By this time he had quite an audience, most of whom were open-mouthed in amazement, and of whom Toady was totally oblivious, concentrating as he was on the job in hand – literally. After he closed the lid of the box a burst of applause rang out and Toady, being a modest, self-deprecating sort, merely blushed, bowed and smiled. Actually, I think it was more like a grin of satisfaction, knowing that he had added another colony of bees to his apiary.

Nepeta sp. (Catmint)

Different types of honeybee

From what I have been saying you may have got the impression that you might encounter only one type of honeybee, but just as there are different species of bumblebee there are different honeybees spread across the world. Ten different species of honeybee have been identified but the only honeybee native to Europe is *Apis mellifera* or the Western honeybee. Within this species there are 25 subspecies, or races, including: *Apis mellifera mellifera* (the dark European bee); *Apis m. ligustica* (the Italian bee); and *Apis m. carnica* (the Carniolan bee, originally from Austria and the Balkan regions). They all have distinctive characteristics and traits.

Apis mellifera mellifera is a rich, dark brown, stocky bee, well suited to cooler northern climates. She is regarded as a good worker and fairly gentle to handle.

Apis mellifera ligustica originated in Italy and Sicily. She is a gentle bee whose colour can range from a dark leather colour to golden yellow or even very pale yellow. This is the most widely distributed variety in the world although it is less hardy in cooler regions.

Apis mellifera carnica, too, is gentle but is generally dusky brown-grey in colour with stripes of a lighter brown. This is the second most popular variety.

Bees have interbred over the years and many would argue that there are very few 'pure bred' honeybees to be found nowadays. So if you do see a honeybee in your garden it is bound to be an *Apis mellifera* of some description.

Crossing bees

Sometimes bee-keepers will, for various reasons, cross bees themselves. We keep bees in the north-west of England, which can be tricky. The weather is often inclement, to say the least, so we need a bee that is robust enough to withstand the climate, but gentle enough to be handled. We tried crossing a dark European bee bred in Scotland (bound to be able to cope with Lancashire weather) with the Caucasian bee (*Apis m. caucasica*), known for its gentle nature. We waited with baited breath, expecting a peace-loving, 'flower power' type bee but wearing a raincoat and wellies. What we got was a feisty Frankenstein's monster of a bee, ready to do battle with anything or anyone that even looked at it the wrong way, let alone tried to relieve it of its honey. Needless to say we did not pursue that line, but we are now getting very close to breeding the bee of our dreams.

Hydrangea aspera 'Macrophylla'

The colony

A colony of honeybees is made up of three 'castes' of bees: the queen, of which there is usually only one; the males or drones, which range in number from a handful, at most, during the winter to a few hundred at the height of the season; and the sterile females or workers, which can number as few as 5,000 or as many as 60,000. Why do the numbers fluctuate so widely? Fundamentally this is due to their roles within the colony. First, let's consider the queen herself.

The queen

Simply put, the queen is an egg-laying machine. Before she can lay eggs, however, she has to mate. When a new queen emerges from her cell she will fly from the hive to an area where mating takes place (known as the drone congregation area). Here she will mate with up to 30 drones, which helps to prevent inbreeding and encourages genetic diversity, and will then return to the hive to start laying eggs. In her prime she can lay up to 2,000 eggs a day.

Her lifespan is about four years and during that time she will remain in the hive unless a new queen emerges, in which case she (the 'old' queen) leaves in a swarm to find a new home, or she is replaced by the bee-keeper. So you will not see her buzzing around the garden. The queen is entirely dependent on the worker bees to feed and groom her and to keep the colony at the optimum temperature and humidity for her to lay her eggs and for them to survive. Without the attendance of the workers the queen, and the colony, would die.

The drones

The drone, as we have seen, mates with the queen. It is a vital role, but that's all he does. Drones do not contribute to the colony in any other way – I have heard them described as the equivalent of couch potatoes, lazing around on the settee all day being supplied with lager and takeaway food. Sylvia Plath, the celebrated poet, describes them as: 'blunt, clumsy stumblers, the boors'. You get the picture! Which is probably why, once he has done his job, he dies, literally on the spot – or, I should say, in the air. Those who are not successful in their attempt to mate are rarely, if ever, allowed back in the hive and they perish. Indeed, the only time he leaves the hive is to mate with the queen or if he is driven out by the workers, so the chances of seeing a drone in the garden are nil. During autumn and winter, or at other times during the season when there is a lack of food, or circumstances prevent the colony from thriving and survival instinct takes over, the workers will kill the drones or eject them from the colony. They will even destroy drone brood, or 'baby' drones. There is no need to nurture vast numbers of drones because their task is so limited.

The workers

It is the workers, or sterile females, who undertake all the jobs inside and outside the hive. This is why there are so many of them, especially during the height of the season. Depending on her age, the worker will carry out tasks such as cleaning cells, tending the brood and the queen, building the comb and dealing with food brought into the hive, and guarding the hive. One of the most important jobs, however, is foraging for food. This is needed not only to feed the queen, babies, drones and other workers immediately, but to lay down as stores (honey) for the winter months. A worker's lifespan will vary depending on the time of year: during the height of the season, when she is most active, she will live for between 15 and 38 days; during the winter, when she is less active and her task is to ensure the survival of the queen, she may live for more than 20 weeks. Worker honeybees are the only ones you will come across in the garden.

────────────── *Two patterns emerge* ──────────────

At this stage we can see two interrelated patterns emerging. First, there is an inverse correlation between lifespan of the worker bee and size of the colony: the lifespan of the individual worker during the summer is quite short but there is a huge workforce to carry out the tasks; during the winter, when she lives longer, a much smaller number of bees is required.

Second there is a pattern that is entirely dependent on the availability of food. During the summer months, when there is at least sufficient, if not abundant, food, the number of bees within a colony is at its height. During the winter months, when there is little or no natural food and the bees are living off stored resources, the number of bees is at its lowest.

It is vital for the survival of the colony, however, that there are sufficient worker bees at the beginning of winter to keep the queen alive. When the temperature drops below 18°C, a good number of bees will start to move together in the hive to form a ball, or cluster. If the temperature falls below 13°C, all the bees will be clustered together to keep the centre of the ball, where the queen is, at a constant temperature of about 35°C. In order to maintain such a degree of warmth, the bees need food.

What we, as gardeners, can do is to provide as many food-bearing plants as we can to help the colony build and produce enough food to sustain themselves throughout all the seasons.

Pollination

One reason for keeping bees is that they produce honey, and indeed that is the main reason we keep them. Arguably, of even greater importance than making honey, however, is the role the honeybee plays in the pollination of crops. Because honeybee hives can be transported, the bees can be moved to wherever a crop is in need of pollination. The efficiency of the bee means that fruit and vegetable yields can be increased by anything up to 36% if they are active on the crop. This is a huge advantage for farmers and growers. (And if it is true for commercial crops then it is also true for domestic ones.) Estimates of the current annual value of UK agricultural pollination by honeybees varies from £200 million to £440 million; in the USA it is estimated to be a whopping $24 billion. Another much-quoted statistic is that a third of all we eat depends on the honeybee for pollination. As well as pollinating commercial crops, it is also thought that honeybees provide pollination for more than 50% of wild plants on which the greater part of other wildlife eventually depends.

The reduction in the number of bees can have a significant economic impact on global food production. For example, the United States Department of Agriculture reported that, in 2007, the cost to farmers of reduced pollination services due to the collapse of bee colonies was $15 billion. According to a study for the United Nations Environment Programme, entitled *The Economics of Ecosystems and Biodiversity*, the worldwide cost in the same year amounted to some $190 billion – a not insignificant sum.

What happens when there are no bees?

To get a glimpse of what would be necessary if there were no bees at all to pollinate crops, we can look to the Sichuan province of China. In the 1980s the number of pear orchards increased, especially with the introduction of a new variety, the fruit of which fetched a higher price than other varieties. Unfortunately, at about the same time, there was a serious outbreak of pear lice (*Psylla*), a pest which is really bad news for pear growers. (*Psylla* can weaken the tree and reduce both growth and fruit bud set. It can also cause premature leaf drop.) The trees were sprayed intensively with an insecticide to eradicate the bugs. Spraying was carried out so vigorously that entire orchards would be treated as many as 12 times during the production season. The pear lice were successfully eliminated – but so were any other insects present, including honeybees which were working the blossom. Bee-keepers removed the remaining bees out of the area in order to protect what colonies were left. Pear blossom is now pollinated, not by honeybees, but by 'human bees', using 'pollination sticks' which are dipped into containers of pollen and then stroked onto each blossom in turn. It is labour-intensive – and, it could be argued, totally unnecessary.

Moving bees

It is at this point, however, that I must utter a word of warning. Bees do not like being moved. If you are an amateur bee-keeper with only one or a few hives at most, then the chances are that you would not want to move them anyway. At the other end of the scale is the commercial bee-keeper, who by necessity has to move them. In America bees are routinely transported from one coast to the other, following the flowering times of vast acreages of commercial fruit and vegetable crops. In Britain this is not necessary. Either way, sensitive (and sensible) commercial bee-keepers will not move their charges any more than is necessary. The bees that I help to take care of are moved a maximum of four times during the season, and then only a distance of less than 30 miles each time. If you do move your bees it has to be less than a couple of feet or more than three miles (the furthest distance which bees will fly from the hive to find food) otherwise their internal 'satnav' will be jiggered.

Stress

Moving bees causes them stress. High levels of stress, and the subsequent possible inability of a colony to combat diseases or to withstand pests, have been posited as a contributory factor to Colony Collapse Disorder (CCD). (In a nutshell, this is where bees leave the hive and do not return.) Recent research in America has shown that many bees collected from collapsed colonies were infected with both a virus called invertebrate iridescent virus (IIV) and a fungus, *Nosema apis*. Both infect the bees through the gut, so researchers are led to believe that the bees' nutrition may also play a part. (In countries where CCD is prevalent, the bees often feed on one type of crop at a time and are then moved on to another. As an extreme, and perhaps not entirely analogous, example, imagine, as a human, eating nothing but fish and chips for a month, and then having apple pie for three weeks solid. You get the idea.) What is not known is how the virus (IIV) and the fungus (*Nosema apis*) interact, but the combination of the two appears to be fatal. It is thought that the weakened bees' complex navigation system is affected and they are unable to find their way back to the hive. Fortunately, at the time of writing, no incidents have been reported in Britain.

What do bees need?

But enough doom and gloom. The bottom line is – we need bees. But what do bees need? Like every other living creature on our planet, honeybees need food and water. Water is water whatever and wherever you are (we shall look at bees' water requirements shortly), but food is a different story entirely. Humans are omnivores, living happily on an extremely varied diet. Some creatures are herbivores, eating only plant-based food, while others are carnivores, eating only meat. Bees are herbivores and their food, and the source of all their nutrition, is specialized, consisting of only two items: pollen and nectar. So if we want to help our bee population it is useful to know a little about their food and how they source it.

Bees need plants

Bees and flowers are interdependent: they rely on one another for their very existence. The bee–flower relationship has evolved over millions of years, with flowers in particular adapting to attract pollinating insects. Bees are active from early spring, usually March, until the first frosts. These often occur in October but sometimes not until November, or even later, especially in towns and cities where the temperature is often 2–3°C – and can be as much as 6°C – higher than in the countryside. (This is a recognized phenomenon known as the 'Urban Heat Island Effect', first described as long ago as the early 1800s.) This period of bee activity corresponds with the main growing and flowering season of plants, which is not merely a coincidence since bees require a substantial amount of food in the form of pollen and nectar from flowers to survive.

Bees need pollen

Pollen is produced by the male sex organs of the plant and, in order for the plant to reproduce, the pollen must be transferred from the stamen to the stigma, the receptive female organs (Figure 3). The most efficient way for the plant to bring this about is either to be self-fertile or to use an agent, either in the form of the wind or by means of an insect. Many plants have evolved to take advantage of wind pollination (particularly grasses) but it is insect-, and, in particular, bee-pollinated flowers which are of interest to us.

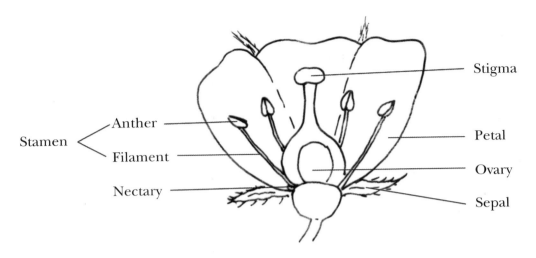

Figure 3 Simplified cross section of an 'open' flower

───────── *Pollen equals protein* ─────────

Pollen provides bees with an essential foodstuff that all living creatures need in their diet for the growth and repair of body tissue, namely protein. The protein content of different pollen can vary considerably – some is as little as 2%, yet it can be as much as 28%. There are three particular reasons why bees need the protein found in pollen.

First, pollen is a vital component of the food that is fed to worker bees from the third day of their lives as larvae. (During the first two days they are fed a special 'brood food' or royal jelly, a secretion from the nurse bees which is in some ways the equivalent of colostrum that mammal babies receive from their mothers.) They are then fed a mixture of royal jelly, pollen and honey until the cell is sealed and they start to pupate. Drone larvae receive a slightly different combination of food, and for slightly longer than the worker larvae, and a queen larva is fed exclusively on royal jelly (Figure 4).

Secondly, it is vital for the sound development of the hypopharyngeal gland (Figure 5). Without getting too technical, this is a gland situated in the head of the worker bee which is crucial for the production of brood food that is fed to larvae during the first two days. The hypopharyngeal gland is also responsible for the secretion of wax with which to build combs, ‖ No. and for the conversion of nectar into honey. Pollen which is fed to young bees within the first five to six days after they have emerged from the cell

Wax glands produce wax.

15

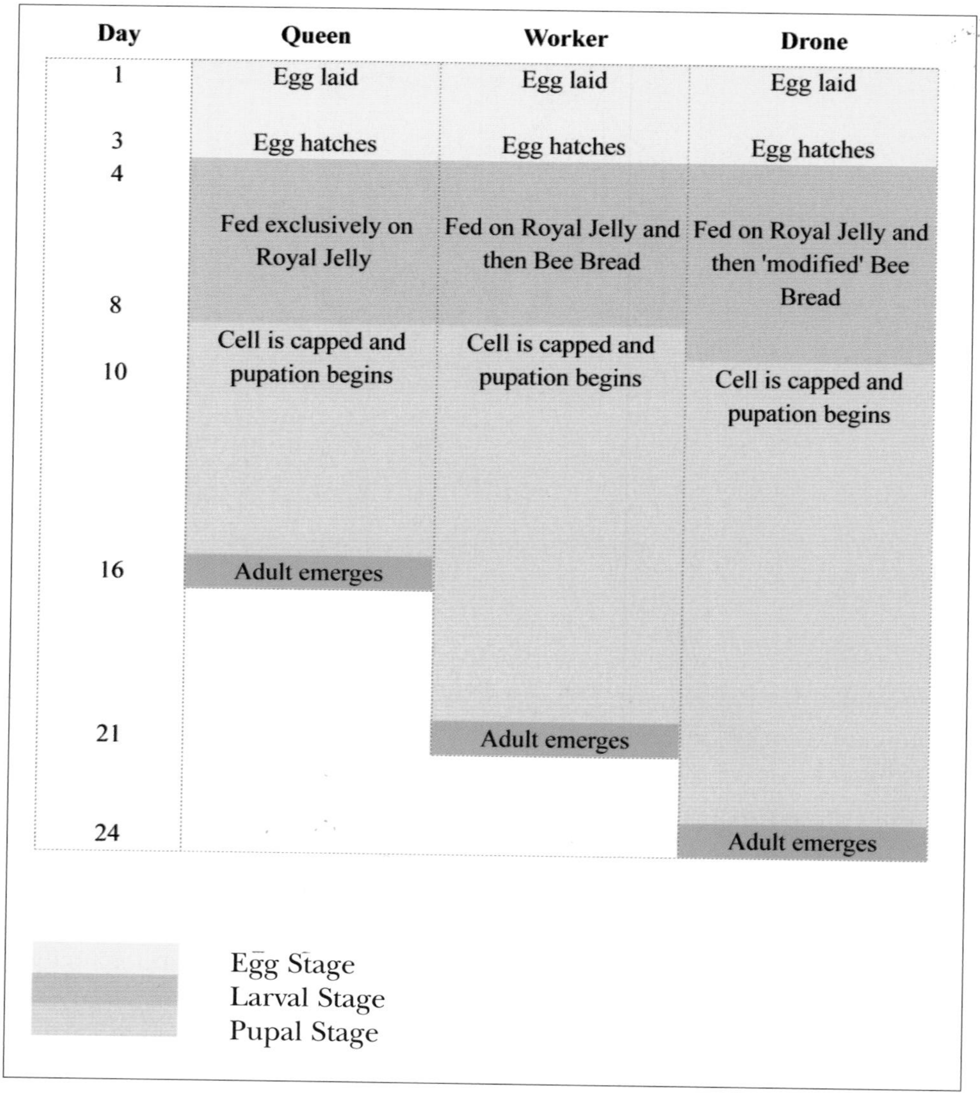

Day	Queen	Worker	Drone
1	Egg laid	Egg laid	Egg laid
3	Egg hatches	Egg hatches	Egg hatches
4			
	Fed exclusively on Royal Jelly	Fed on Royal Jelly and then Bee Bread	Fed on Royal Jelly and then 'modified' Bee Bread
8			
10	Cell is capped and pupation begins	Cell is capped and pupation begins	Cell is capped and pupation begins
16	Adult emerges		
21		Adult emerges	
24			Adult emerges

Egg Stage
Larval Stage
Pupal Stage

Figure 4 Honeybee development

ensures that the hypopharyngeal gland grows properly.

Third, the proper development of bees' fat bodies, or fat cells, depends on adequate protein intake. Fat bodies, which store not only fat but also protein and glycogen ('stored' glucose), can be found in the abdomen of the bee. These fat bodies are particularly important in bees which must overwinter: not only must they survive the winter, but they must also be fit

External

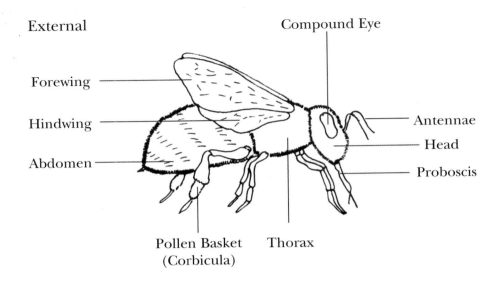

Compound Eye

Forewing

Hindwing

Abdomen

Antennae

Head

Proboscis

Pollen Basket
(Corbicula)

Thorax

Internal

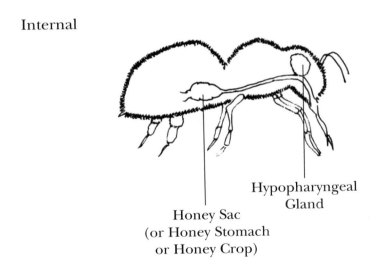

Hypopharyngeal
Gland

Honey Sac
(or Honey Stomach
or Honey Crop)

Figure 5 Honeybee anatomy

enough to be able to provide food for the brood, or baby bees, early the following season.

———————— *How much pollen do bees need?* ————————

It has been calculated that between 100 and 125mg of pollen is needed to rear one worker bee so, if during the summer a colony rears between

150,000 and 200,000 bees, up to 25kg of pollen is required. Bearing in mind that a typical pollen load weighs between 10 and 30mg, it means that up to two and a half million pollen forages must be made.

Collecting pollen

Not all forager bees collect pollen, however – this work is left to a maximum of 25% of the foraging population. Pollen foragers will generally fly from one flower to another of the same species until that resource is exhausted (although between 1 and 10% of pollen samples have been found to contain pollen from more than one species). This behaviour is called constancy. Constancy is important for successful pollination as the flower must have pollen from the same species in order to be fertilized. It also reduces the distance that bees have to fly to collect a full load.

Although individual bees will remain constant to one pollen or nectar source, collectively they will visit a range of sources within the foraging zone. This was brought home to me when I was helping to check the bees that we had placed on an agricultural rape crop which covered many acres. Rape pollen is generally a pale yellow colour and this is what I expected to find in the returning bees' *corbicula* or pollen 'baskets'. While this was indeed the case for the majority of bees, I also saw that some bees had collected pollen of an almost burgundy-red colour; others had dark blue, almost navy pollen; still more had yellow-orange pollen (page 19). It turned out that, adjacent to the field where the bees were working, there was another rape field which had failed and had been taken over by *Lamium purpureum* (red dead nettle), which does bear burgundy-red pollen. The yellow–orange pollen had come from the road verges which were almost a solid blanket of *Taraxacum officinale* (dandelion). It seems that bees like a varied diet too!

Pollen release

Another important factor to note is that plants allow pollen to be released at different times of the day and over varying periods of time during their flowering season. For example, pollen from individual florets on the head of *Taraxacum officinale* (dandelion) will be made available for about 10 minutes, mainly during the morning; whereas pollen from *Rubus fruticosus* (bramble) (page 23) can be collected from individual flowers for up to four days, throughout the day. Although I have said that forager bees tend to stay

Pollen in the comb

loyal to one species until the food source has run out, this behaviour has been known to change if different supplies are available at different times of day; for example, a bee will collect from *Papaver rhoeas* (field poppy) in the morning and broad bean in the afternoon because this is when each flower releases its respective pollen.

Pollen colour

Pollen colour varies, too. Most pollen is a shade of yellow or orange, but it can range from the clotted cream colour of *Cichorium intybus* (chicory) to the dark purple, almost black of *Papaver orientale* (Oriental poppy) and most shades and hues in between (Figure 6).

What happens to pollen in the hive?

When pollen is brought back to the hive it is packed into cells to exclude air, and a little regurgitated honey, hypopharyngeal secretions and enzymes are mixed in with it to hinder decomposition and fermentation, and to aid initial digestion. This pollen with 'additives' is known as 'bee bread'. It is stored in cells close to the brood cells so that it is readily available to the nurse bees feeding the larvae and to the newly emerged adult bees.

Bees need nectar

Nectaries

Nectar is the sugary fluid that is secreted by plant nectaries. Nectaries can be categorized as either 'floral', which means they are found within the flower, or 'extra-floral', which are located outside the flower – usually, but not always, at the point where a leaf joins the stem. For pollen-bearing plants, floral nectaries are of the greatest importance because the presence of nectar attracts pollinating insects to the flower. Floral nectaries are usually found at the base of the ovary, which means that the insect has to brush past the pollen-bearing stamens to get to the nectar. In effect, nectar is the insects' reward for being 'hijacked' by the flower to distribute its pollen.

The amount and frequency of nectar that is produced varies from flower species to flower species. Even within each species the amount of nectar secreted can fluctuate enormously depending on factors such as the robustness of the plant and the age of the flower. Environmental issues also play a part, such as temperature and humidity.

What is nectar?

Nectar is the main source of carbohydrate for bees. It is basically liquid sugar in the form of sucrose (a disaccharide), glucose and fructose (both monosaccharides), in varying amounts. Nectar also contains vitamins (especially vitamin C and some in the vitamin B complex), minerals (such as potassium and calcium), amino acids and other substances.

Is all nectar the same?

There are basically two types of nectar. The first contains chiefly fructose and glucose, with fructose making up the greater part. This nectar is generally associated with open or cup-shaped flowers which, for honeybees, are easy to reach (page 9). Obvious examples of accessible nectaries are those of the Rosaceae family, such as *Fragaria* (strawberry) (page 21), *Rubus fruticosus*

Figure 6
Pollen colour of selected plants

Colours are approximations and are for illustration only

Achillea millefolium

Aesculus hippocastanum

Aster novi-belgii

Borago officinalis

Centaurea cyanus

Cichorium intybus

Echium vulgare

Hederea helix

Hyssopus officinalis

Lamium purpureum

Liatris spicata

Papaver orientale

Salix caprea

Sorbus aucuparia

Taraxacum officinale

Fragaria x *ananassa*
(strawberry)

(blackberry) (page 23) or *Crataegus monogyna* (hawthorn) where the single flower is open and nectar is available from the entire inner surface of the receptacle, the apex of the flower stalk in the centre of the flower.

Second, there is nectar whose sugar content is predominantly sucrose; this type of nectar is usually found in tube-shaped flowers (Figure 7). The nectaries of these flowers may not be as easy to get to for honeybees, especially if the nectar is in a reservoir at the bottom of a corolla tube (a group of petals forming a tube).

Research has shown that bees prefer nectar where the ratio of each of the sugars, namely sucrose, glucose and fructose, is the same, but this ratio seldom occurs in nature: here bees usually target nectar with a ratio of 2:1:1. (Interestingly, the higher the fructose content, the longer the resulting honey will remain liquid.) The sugar content of nectar varies between 3% and 80%: for example, *Borago officinalis* (borage) (page 24) nectar contains about 25%; *Trifolium repens* (white clover) (page 25), up to 40%; *Origanum vulgare* (marjoram) (page 25) reaches almost 80%.

How much nectar do bees need?

Each colony needs about 120kg of nectar a year: 70kg is consumed during summer months as immediate food and for energy requirements; the balance is converted into approximately 20kg honey to be used as store food during the winter months (unless the bee-keeper comes along and takes some of the honey, in which case this is replaced by 'artificial' food in the form of a specially formulated syrup).

What happens to nectar in the hive?

Nectar has to be processed before it is turned into honey. During processing the sucrose (diasaccharide) content of the nectar is converted into glucose and fructose (monosaccharides). This is a chemical process which is partly carried out inside the forager bee's honey sac (or honey stomach, or honey crop) during her flight from the foraging site to the hive. (The honey sac is used to store and partly process nectar, rather than to digest pollen and nectar, a process which is carried out in a separate 'proper' stomach. The honey sac is also used to carry water back to the hive.) When the forager bee reaches the hive the nectar is passed to a 'receiver' bee who ingests it, swallows it and regurgitates it many times, and so the conversion process continues.

Rubus fruticosus (blackberry)

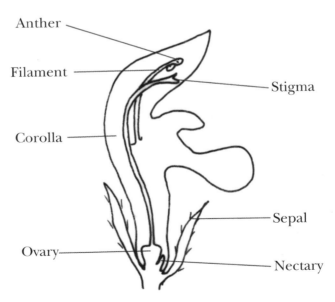

Anther

Filament

Corolla

Stigma

Ovary

Sepal

Nectary

Figure 7 Simplified cross section of a plant with corolla

Borago officinalis (borage or starflower)

Trifolium repens (white clover)
Origanum vulgare (marjoram)

Each time the nectar is regurgitated, the water content decreases. When the sugar conversion is complete the droplet of nectar is placed in a cell in the comb along with other droplets. At this stage the moisture content may still be too high, so other worker bees within the hive will fan their wings to aid evaporation until the water content is below 19%. This is the point at which naturally occurring yeasts in the honey cannot grow and therefore the honey will not ferment. The cell is now sealed with a wax cap. This capped honey is what is known as mature or ripe honey, and this is what the bee-keeper harvests from the hive.

Collecting nectar

A typical nectar load is between 25 and 40mg, so, as we saw with the pollen, a huge number of nectar-foraging flights have to be made. The majority of foraging bees collect nectar, although 17% will collect both pollen and nectar. It has also been reported that bees will change their allegiance from a pollen crop to a nectar crop but never the other way round.

'Hazardous' nectar

Some nectar, however, is, if not poisonous, then hazardous to bees. Fortunately in Britain there are very few plants which yield such nectar, but be aware of some varieties of rhododendrons and azaleas (particularly *Rhododendron ponticum*), and the *non-native* limes *Tilia petiolaris*, *T. oliveri* and *T. orbicularis*. Although the nectar of *Senecio jacobaea* (ragwort) and *Ligustrum ovalifolium* and *L. vulgare* (privet) is not perilous to bees, it produces honey which is unpalatable to humans. Ragwort honey is a deep yellow in colour and has a strong flavour and aroma; privet honey is dark, quite thick and has a bitter flavour.

Bees need water

Like all living creatures bees need water for their own metabolism. It is also vital for the welfare of the colony as a whole. In order for the brood (baby bees) to develop properly the temperature within the brood area must be kept at 35°C ±0.5°C (although the brood can survive between 32 and 36°C), and the relative humidity has to be between 35 and 45%. If the temperature within the hive reaches 47°C then the comb will collapse, spelling disaster for the colony.

Bee 'air con'

In order for the optimum temperature and humidity to be maintained, the bees instigate an 'air-conditioning' system, bringing water into the hive which then evaporates, keeping the temperature down and the humidity stable. There are occasions when there is too much moisture in the hive, however. This usually happens when there is a high water content in the nectar which is being collected. Again, bee 'air conditioning' is switched on – some bees will fan their wings near the entrance of the hive to increase the air circulation and aid evaporation.

How much water do bees need?

Water is also a constituent part of the food that is given to the brood. Unlike pollen, which can be stored in cells, water has to be collected as required. At the start of the season, in particular, water is also needed to reconstitute the store honey; bees do not 'eat' honey in its concentrated form – it has to be diluted. All of these factors mean that the amount of water a colony may need at any time can vary considerably. Estimates fluctuate between 120ml and 4 litres per day, although it has been reported that in Australia during a heatwave, a 90 litre tub of water was emptied in less than three hours by bees from 150 hives. To give you an idea of just what a feat that was, one bee can carry approximately 25mg (this is equal to 0.025ml – about 1/200 of a teaspoon) of water at a time!

Access to water

Although bees will fly considerable distances to find water it is better for them (if you keep bees yourself) if you can provide some close to the hive. A pond with a pebbled edge is a luxury which you do not need to aspire to, although it will attract no end of wildlife to your garden. Bees will naturally take water from wet surfaces such as pebbles, soil, grass and even wet washing, rather than from an open surface. They cannot land on water without breaking the meniscus (the 'taut' surface of the water), and since they are not good swimmers they need a good solid surface to land and walk on in order to access the water. In addition, as we have seen, in warm weather a strong colony will require several pints of water a day, so a good, clean, regular supply is essential. Any watertight container filled with pebbles or crocks and topped up with clean water is ideal but I have found that a poultry feeder with some pebbles in the canal at the bottom where the chickens would normally drink from is an excellent solution. Do not site it too close to where you sit or where you regularly walk, but put it far enough away from you, and close enough to the hive, to encourage the bees to use it as their watering hole. A word of warning though: as far as collecting water is concerned bees are creatures of habit – once they detect and adopt a source of water it is very difficult to persuade them to go elsewhere, even if a new supply is placed closer to the hive.

Let's now look at the aspects of the bees' anatomy which allows them to detect, access and collect pollen and nectar, and also how they communicate the location of the food to others in the hive.

Bees' vision

Bees 'see' differently from us. Humans can see all the colours of the rainbow from red through to violet (750–370 nanometres (nm)). Bees' vision is slightly different in that they see less than we do at one end of the colour spectrum but more than we do at the other (650–300 nm) (Figure 8). Their vision is complex and although they can detect a fairly broad range of colours, what they see falls into fewer categories than ours. Bees cannot see red. Because of the difference in light detection, what appears to humans as, say, the white of the wild cherry blossom actually appears to the bee as blue-green. This has a direct influence on which flowers bees are attracted to – they will automatically target flowers which stand out to them. This does not mean, however, that all flowers that appear to us to be outside the bees' visual spectrum are not appealing to them. Think of *Papaver rhoeas* (field poppy) – probably one of the most vivid reds (as we see it) in the flower world. But the red that we see also contains pigments which reflect ultraviolet. And if we look at the centre of the flower, where the 'business area' is located, we see that there are dark blotches. These, along with the UV pigments, are the bits that the bee 'sees' and in fact the pollen of the field poppy is dark blue, which is well within their field of vision.

Movement detection

Bees are also able to detect movements at a much faster rate than humans. Bees' eyes are compound eyes, made up of thousands of tiny lenses which collect images which are then combined by the brain into one large picture.

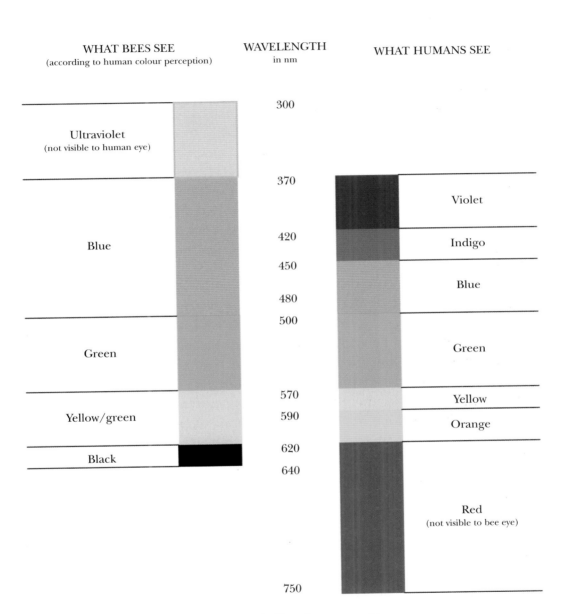

Figure 8 Visual spectrum of bees

They can also easily distinguish between solid and broken patterns, with a preference for the latter. So a mass of *Rudbeckia* (coneflower) (page 31), waving slightly in a gentle breeze, will be an easy target for the honeybee and when they do start to forage, they do not have to move very far from one pollen and nectar source to another.

Scent detection

Bees can also detect scent by using their antennae. These have odour receptors to identify minute scent molecules carried in the air. We do not know if bees analyse fragrance in the same way that humans do (to us, a rose has a pleasant smell, but rotten eggs have not!) but research has shown that a bee's sense of smell is so exact that it can distinguish between hundreds of diverse aromas and can also tell whether a flower has pollen or nectar, by sniffing its scent from several metres away.

The pollen basket

Bees are covered in hair, which allows pollen to 'stick' to the bee's body. Some of the pollen which is dispersed over the body is transferred from one flower to another, facilitating pollination. Later, in flight, the rest of the pollen is collected together through grooming movements of the front and middle legs to be compressed into the *corbicula* or pollen 'basket' on the hind legs. The *corbicula* is not actually a basket but an indent fringed by hairs which holds the pollen in place. I have also observed bees coming back to the hive with pellets of the almost white pollen of *Impatiens glandulifera* (Himalayan balsam) sticking to the back of the thorax – whether this was by accident or design I'm not sure, but there were quite a number of workers with this 'extra load'.

The size, shape and surface structure of pollen grains can vary considerably. For example, the pollen grain of *Myosotis* (forget-me-not) measures 7 microns (a micron is one millionth of a metre) in diameter, whereas the pollen grain of *Cucurbita pepo* (squash) is some 200 microns. These differences almost certainly affect collection by bees. During a time of surfeit bees will undoubtedly collect pollen that is not only easy to access but also easier to manipulate and carry.

The proboscis

Bees collect nectar and water by means of the *proboscis*, which is in effect a long, slender, hairy tongue which acts as a straw, drawing up nectar or water. When it is being used, the proboscis moves rapidly back and forth while the supple tip performs a lapping motion. After feeding, the proboscis is drawn up and folded behind the head.

Does size matter?

When trying to access nectar, the length of the bee's proboscis is extremely important. In the honeybee (*Apis mellifera*) the length of the proboscis can

Rudbeckia sp. (black-eyed susan)

range from 5.7mm to roughly 6.8mm, depending on the species of bee. Bumblebees (*Bombus* sp.), however, have a proboscis length of 7.2mm to 13.5mm, depending on the species. In order to illustrate why this is important, let us look at *Trifolium pratense* (red clover). The corolla tube (the tube which is formed when petals 'fuse' together) may be 7.5mm to 12.4mm long. Nectar is secreted at the base of the corolla tube but only extends 1.35mm to 1.47mm up the tube. So a bee needs a minimum proboscis length of 6mm in order to access the nectar, effectively barring a good number of honeybees. (All is not lost, however! If red clover is cut once or twice the subsequent flushes of flowers will have shorter tubes which the honeybee can reach.) *Trifolium repens* (white clover) (page 25), however, has a corolla tube-length of roughly 4mm, making the nectar accessible to more bees. (See Figure 9 for more examples.)

Honeybees		
Apis mellifera mellifera	5.7 – 6.4	
A. m. ligustica	6.3 - 6.6	
A. m. carnica	6.4 - 6.8	
Bumblebees		
Bombus lucroum	7.2	
B. terrestris	8.2	
B. pascuorum	8.6	
Average depth of corolla tube	mm	
Rubus fruticosus	0.0	
Melilotus officinalis	1.7	
Borago officinalis	1.8	
Trifolium repens	2.5	
Onobrychis viciifolia	4.3	
Knautia arvensis	6.0	
Lavandula angustifolia	6.0	
Echium vulgare	6.7	
Salvia uliginosa	7.0	
Trifolium pratense	7.5	
Trachelospermum jasminoides	8.1	

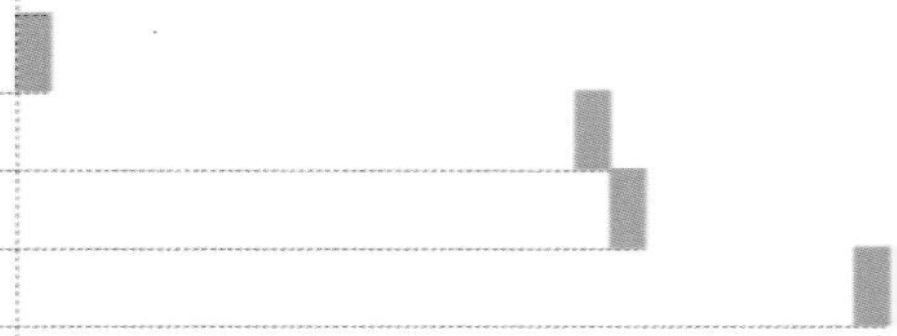

Figure 9 Proboscis and corolla length

The waggle dance

Once a worker bee has found a source of nectar, pollen or water she will return to the hive to tell the other worker bees. Bees do not have a system of language as we know it, but they do have an ingenious method of communicating. This is the famous 'waggle dance'. The dance is a figure-of-eight movement which takes place on the vertical surface of the honeycomb. The worker bee starts out in a straight line and waggles from side to side. She will then circle round to the beginning and waggle along the straight line again, but this time she will circle the opposite way back to the beginning, completing the figure of eight. She will carry on doing this until she has communicated all the information. Research has shown that the duration of the straight line 'waggle' phase depends on how far away from the hive the food source is, and the number of times she completes

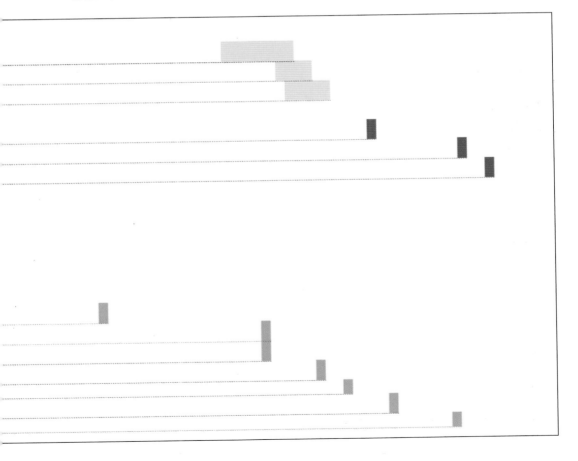

the figure of eight shows how much food there is. In addition, the angle of the straight line from the vertical corresponds to the angle between the food source and the sun, thereby showing the direction in which the bees should fly when they leave the hive. Truly amazing.

How do plants attract bees?

It is true that not all plants are pollinated by bees (some are pollinated by other insects, some are wind-pollinated, some reproduce vegetatively), but for those that are, the relationship between bees and plants is symbiotic: no bees, no plants; no plants, no bees. But in order to attract bees, the plants must advertise themselves; they must 'sell their wares', so to speak. If they do not, it is like a restaurant serving first-class food but having done nothing to tell the public that they are there – no sign, no advertising, no menu displayed equals no customers. The main means by which plants advertise is through colour and scent so if, through these two attractants, they can make themselves as alluring as possible to the pollinating insect which suits them best, then they have a far better chance of reproducing. It is interesting to note that while the colours of many cultivated varieties of flowers have been bred to be attractive to the human eye, the colours of wild flowers correspond to the colours their respective pollinators can best see.

Nectar guides

Like the poppy with its dark blotches, many other flowers have marks on their petals which act as 'nectar guides'. These are a bit like runway landing lights for aeroplanes: they are the flowers' way of guiding the bee to the source of the nectar. They are frequently darker than, or of a contrasting colour to, the rest of the flower and often take advantage of the bees' ability to see UV light.

I came across a good example of this when I recently visited a private garden open to the public. There were several beds planted up *en masse*, each with a different variety of *Penstemon*. One variety, in particular, was positively humming with different kinds of bees, especially honeybees – I had difficulty finding a stem that did not have a bee working one of the flowers. The flowers had bands of varying hues of pink with white throats, and, of greater interest from a bee's point of view, pink and white stripes leading from the edge of the bloom right into the centre. The nectar and pollen could not have been better advertised if there had been neon lights, a brass band and a town crier all declaiming their presence! (I am reliably informed that this variety was *Penstemon* 'Tubular Bells Rose' (page 36).) Interesting, too, was the fact that in the next bed was a darker, cerise variety (*P.* 'Tubular Bells Red') with very few 'landing stripes' (to my eyes, at least) which, bar a couple of bumblebees, was devoid of any bees at all.

What plants are best for bees?

As we have seen, however, it is not enough for a bee to be attracted to a plant if it cannot access either the pollen or the nectar or both. This is where the nature of the plant itself and the shape and form of the flower have a bearing. Let's look at the nature of the plant first.

The nature of the plant

In a nutshell, the nearer the cultivated plant is to its wild, 'natural' original, the better it is for bees. This is because these plants have evolved naturally, without human intervention, to attract the pollinating insects best suited to their needs. I have an area beside our garage that I have given over to British native perennial and annual flowering plants: it is a patchwork of colours from spring through to autumn and it attracts no end of insects, especially bees. My neighbours (with whom I get on very well) call it my weed patch. It is all in the definition. One common definition of a weed is that it is a plant growing in the wrong place. Well, my 'weeds' are growing in the right place; I have put them there on purpose, so they must be plants. I admit (sometimes grudgingly) that there may be a few specimens that under any other circumstance would be deemed weeds, such as *Taraxacum officinale* (dandelion) (which I never allow to set seed, by the way), and *Ranunculus acris* (meadow buttercup), but there are also beautiful wild flowers, such as *Leucanthemum vulgare* (ox-eye daisy) (page 37), *Centaurea cyanus* (cornflower) (page 39) and *Succisa pratensis* (devil's bit scabious). It is true that sometimes my patch looks as if it is having a 'bad hair day', but what it lacks in attractive order and structure, it makes up for in its attractiveness to insects. Incidentally, I have discovered an ingenious way of making your wayward patch look as if it is tended (and intended!) and has not just been left to its own devices: lay a strip of turf around its perimeter and keep this well mown – anything growing within its bounds automatically looks 'cultivated'.

Penstemon (Tubular Bells Rose)

Leucanthemum vulgare (oxeye daisy)

The shape and form of the flower

The shape and form of the flower is also important. It is beyond the scope of this book to look into the botanical detail of plants and their different flower structures but there are some hints and guidelines that are worth considering when it comes to the shapes and forms of flowers which are of benefit to honeybees.

Single or double flowers?

Should we grow plants with double or single flowers? The short answer is, single flowers, every time. Highly-bred or hybridized plants often have double flowers which, although they look good from a human perspective, are of little or no value to bees. This is because the 'extra' petals of a double flower are in fact a genetic mutation of the sexual structures of the flower, which means that there are no, or very few, pollen-bearing male stamens, rendering them sterile, and often devoid of nectaries. Others may have so many petals that even if nectar and pollen are present, the bees cannot get to it. It is actually possible to visualize a swathe of colourful, fragrant flowers in which the honeybee would starve. Double flowers are fairly easy to spot: a modern rose, a showy delphinium or even a double-flowered wisteria. And anything with *flora pleno* in its Latin botanical name will automatically alert you to the fact that it is double-flowered.

Can bees reach the nectar?

There are some beautiful plants where the nectar is out of reach of the honeybee, although it can be accessed by bumblebees and butterflies. This is the case with flowers where the nectar is so deep-seated in the corolla tubes that the honeybee cannot access it, as we have seen with clovers, for example. There are also those flowers which have evolved to allow access by some insects but not others. A good example of this is *Antirrhinum majus* (snapdragon) (page 41). It would appear that it is all but impossible for any insect to get to the nectar, it being fully enclosed within the flower. The bumblebee, however, has learnt how to exploit the flower. Using her body weight she manages to push and squeeze her way in so that she can extend her proboscis, or, if necessary, crawl further down the flower, to reach the nectar. At this point she is fully enclosed by the flower – quite scary! But the nectar reward makes it worthwhile. The honeybee, however, is nowhere near heavy enough to prise the flower apart, so access to the nectar is denied.

Centaurea cyanus (cornflower)

'Robbing'

What is interesting, however, is that even though bumblebees can access nectar in such flowers as clover and snapdragons, they will often take a 'short cut' to the nectar, essentially 'robbing' the flower because they do not pollinate it. They do this by edging down the outside of the flower and piercing the base of the corolla tube where they think the nectar is most likely to be situated. They then insert the proboscis and take the nectar. Other insects, especially honeybees, can take advantage of the hole which the bumblebee leaves to access the nectar, which otherwise would be out of reach – this is known as 'secondary robbing'.

Flower seasonality

As we know, bees start to forage in early spring (usually March) and will continue until the autumn (September, but often into October or even November if the weather is clement). There are very few plants that flower all year round. One that does, and is also attractive to bees, is *Ulex* (gorse): there is a saying that when gorse is in flower, kissing is in season – which is to say that it is always in flower! It is, therefore, perhaps one of the best bee plants, even if it does not come in the top five of every gardener's 'must have' shrubs. There are some plants that have a very long flowering season *Lamium* (dead nettle) (page 211) or *Taraxacum officinale* (dandelion), for example) but the majority are seasonal, flowering at particular times of the year. It is, therefore, very important to bear this in mind when thinking about what plants to grow for bees.

Ideally, there should be something in flower throughout the foraging season, although there will inevitably be times when there is less pollen and nectar available than at others. An example of this is what many bee-keepers call the 'June gap' (although this may not necessarily always be in June – it is entirely weather dependent), when nectar from broad-leaved trees and spring flowers has long since gone, but the main flush of summer flowers has not yet opened . Vigilant bee-keepers will provide artificial food for their bees during this time to keep them from starving, but if, through judicious planting, there is natural food available, then so much the better.

Five guidelines for choosing plants

We shall be looking at some plants that are in bloom in each season later in the book and more detailed information of some bee-friendly plants is given in the gazetteer, but, in a nutshell, there are five basic guidelines that will ensure that you can make your garden, or at least part of it, as suitable as possible for our buzzy friends:

Choose flowers with high pollen and/or nectar content;
Choose flowers within the bees' visual spectrum;
Try to have plants in flower throughout the 'bee season';
Consider whether bees can access the flowers;
Choose plants as close to the native species as possible.

So next time there is a gap in your garden and you find yourself browsing

Antirrhinum majus (snapdragon)

around the local garden centre, think of the following mnemonic to help you remember what sort of plant to choose.

Paul's	**P**ollen
New	**N**ectar
Van	**V**isual spectrum
Speeds	**B**ee **S**eason
Along	**A**ccess
Nicely	Close to **N**ative species

2
Creating and Maintaining a Bee-Friendly Garden

Most gardening – whether it is for growing vegetables, providing a pleasant place to sit, indulging in the nurturing of a favourite species of plant, or, as we are doing, offering a garden friendly to bees – involves many shared, 'universal' tasks: ground must be prepared, plants chosen and planted, water applied if necessary, debris cleared away and soil improved, and weeds and pests done away with.

In this chapter we will look at how you as a gardener can create and maintain a bee-friendly garden wherever you live and no matter how large or small your garden. That is not to say that the entire garden has to be given up to our buzzy friends, although there is historical evidence that gardens specifically for bees were created. Gervase Markham, writing in the 17th century, says of the wallflower: 'The Husbandman preserves it most in his Bee-garden, for it is wondrous sweet and affordeth much honey.' The fact that he particularly identifies a 'Bee-garden' must mean that specific areas were reserved for beehives, and presumably for planting flowers attractive to bees. In 1943 the cover of Eleanour Sinclair Rohde's herb catalogue featured a design for a complete bee garden by Margaret Oden (Figure 10). Needless to say, the majority of plants are herbs, but the fact that it is called a bee garden and not a herb garden shows that the emphasis is being placed on bees rather than herbs. (What is interesting from a bee-keeper's point of view is that there are two hives shown on the plan. There is no scale given so we do not know how big the actual garden is. I'm not convinced, however, that it would be big enough to sustain two hives unless there was a good deal of forage available outside the bounds of the garden.)

You may feel inclined to turn your entire plot into a bee garden, but even planting up a small area with bees in mind will attract them, and they will reward you with a soothing buzz as they work their way from flower to flower. What is more, you will have the satisfaction of knowing that not only are you helping one of the most industrious creatures on our beautiful planet, but you will also be working to maintain a healthy ecosystem. And, if you fall in love with bees as I did, you may even want to have a hive (and lovely honey) of your own – but that would involve reading another book! But back to the gardening work in hand.

Organic gardening

If you want to have a bee-friendly plot, probably one of the most important things that I would support is to garden organically. To the commercial grower and farmer this means adhering to strict regulations, laid down by the Soil Association, Organic Farmers and Growers, or other recognized authorization bodies. (Anything that you find advertised or for sale in shops as organic has to bear the certification mark of one of the authorization bodies.) But for the average gardener, like me, the practice of organic gardening usually means working with nature and what she provides, rather than introducing man-made products, such as herbicides, artificial fertilizers and the like.

As far as my practice of gardening organically is concerned, I have been described as fanatical but, as with most things in life, you have to find a way which squares with your conscience, with which you are comfortable, and which is also realistic and feasible. I believe that keeping bees, providing a suitable garden for bees, and organic gardening go hand-in-hand.

The killing fields

As a bee-keeper (and bee-keeping gardener) I am extremely conscious of not using anything ending in '-cide', (from the Latin *caedere*, meaning 'to kill'). I don't know if one of my bees would be killed if I inadvertently sprayed it with herbicide and I am not going to carry out an experiment to find out. I do know, however, that, because of my precious bees, I would never, ever, use any insecticide or pesticide in my own garden. I care deeply about all beneficial insects, especially ladybirds, lacewings, hoverflies and other types of bees – without them our entire ecosystem would collapse.

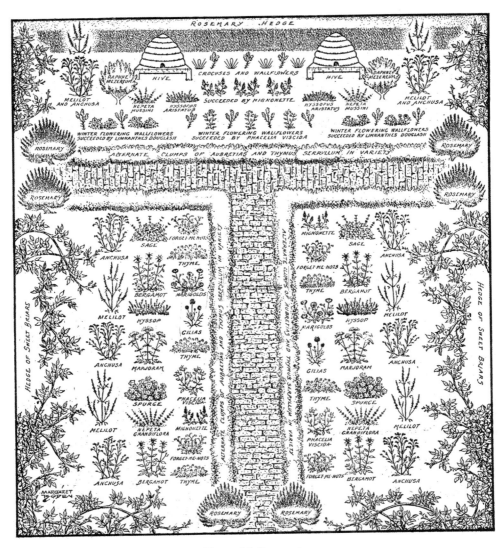

Figure 10 Bee garden

(Wasps are perhaps the exception. Although they do have a role to play in pollination and aphid control, some of them, on occasions, have been known to attack my bees and rob the hive of larvae and honey, so they are definitely insects *non gratae* as far as I am concerned.)

I attempt to maintain a good ecological balance in my garden so that the ladybirds will eat the aphids and birds will keep the caterpillar population down to a reasonable level, so there should be no need to use bug killers anyway.

Soil

The main focus of gardening organically is looking after the soil: it is the lifeblood of anything you plant in it and if you have healthy soil you will have strong, healthy plants, which will be better equipped to withstand the onslaught of any pests or diseases. And from a bee's point of view, the stronger the plant, the more likely it is to reach reproductive maturity, and the more pollen and nectar it is likely to produce. It may be a long-established cliché but, in the words of Arthur Fallowfield, a character on the BBC radio comedy series *Beyond Our Ken*, played by Kenneth Williams: 'The answer lies in the soil.'

In order to keep your soil in tip-top form and supply it with enough nutrients to sustain the plants, it has to be fed and conditioned. By far the best way of doing this is to give it a mulch of compost or manure. (A mulch in this context is a layer of organic matter spread on the soil to provide nutrients, retain moisture and suppress weed growth and seed germination.)

Compost

Compost is basically rotted down organic material. The national charity Garden Organic (see the Useful Addresses section at the end of the book) gives excellent instructions on how to make compost on their website. All garden debris (except woody material and perennial weeds) can be put in a compost bin along with kitchen waste. Waste vegetable and fruit matter (nitrogen-rich materials) can go in the compost bin, along with eggshells, tea bags and coffee grounds, and the occasional bucketful of horse dung if you happen to come across some. I recall the time when I visited my Aunt Mags, who was renowned for having the best roses in her neighbourhood. She lived on the edge of a small town in Sussex and one morning we called by the market to stock up on some lovely locally-reared beef and her favourite 'smelly' cheese. When we returned to the car I went to put our booty in the boot, only to be told that I must put it on the back seat. When I queried this, dear Aunt Mags said it was because of the bucket. It turned out that she kept a lidded bucket in the boot, together with a small coal shovel in case she came across any horse dung on her journeys. Need I say more?

On no account put anything cooked (even vegetables), meat or anything derived from animals, like cheese, in your compost bin – it will attract vermin, especially rats. You should also layer in some dry matter, like shredded newspaper or cardboard (carbon-rich material). The nitrogen-rich and carbon-rich materials will combine and break down, giving lovely rich compost. You can then use your own compost to mulch your garden.

Manure

I cannot make enough of my own compost from vegetable trimmings, dead flowers and the like but I am lucky in that I have a ready supply of well-rotted horse manure available from a local stable. (Manure is basically the droppings of plant-eating animals.) I bribe Toady with as many helpings of tea and scones (with raspberry jam and cream, naturally) as he can decently manage in one sitting, to transport bags of the sweet-smelling stuff for me in his trailer. Mature manure really is sweet-smelling, honestly. You should be able to get hold of a handful, close your eyes and take a jolly good sniff without thinking of ... well, you-know-what. If you can't, then it hasn't rotted down enough.

Other compost

If you can't make your own compost or get hold of any manure, then ordinary, multi-purpose compost that you buy in bags will do a reasonable job, although it won't contain any worms – vital for the well-being of your soil – and it will be a lot more expensive. If you do have to use it, then look out for special offers at garden centres and home improvement stores.

Comfrey fertilizer

If you get the soil right, then there is often no need to add any fertilizer. There are occasions, however, especially with 'hungry' feeders like tomatoes, or if you are growing things in containers, when you have to give the plants a bit of extra sustenance. In this case, I use home-made comfrey fertilizer. This is easy to make but somewhat smelly – no, it is not somewhat smelly, it stinks. The first thing you need is the raw ingredient. I have a small patch of comfrey (*Symphytum* x *uplandicum*) growing in the corner of my garden, the leaves of which I use specifically for making fertilizer. Make sure you get hold of the correct variety, however, as some are very invasive and can take over your garden in no time. The variety you need is 'Bocking 14'. The advantage of this variety is that it does not set viable seed so it will stay where you plant it.

After much experimentation with various Heath Robinson contraptions involving drainpipes and old lemonade bottles, I have found that by far the easiest way to make comfrey fertilizer is simply to get an old watering can – the largest you can find – and cram it with comfrey leaves. Wrap it up in a black plastic bag (a bin liner is perfect) to keep out insects and the rain, and leave it. As the leaves start to rot down, add more leaves and in no time at all a dark liquid will start to accumulate in the bottom of the can. Strain this off into an old plastic bottle, pop the top on, and store it in a dark place until you need it. (Nothing is wasted because you can put the 'mush' – and I use that word on purpose – that is left in the can on the compost heap.)

You cannot use this comfrey liquid 'neat', however. Unlike a good single-malt whisky, which my husband tells me should never be diluted with water, your concentrated comfrey liquid needs to be diluted at a ratio of 1 part liquid to at least 10 parts water, otherwise it will 'burn' the roots of the plants and do them more harm than good.

Other fertilizers

If I don't have enough compost or manure, or have run out of comfrey liquid, I will buy an approved organic fertilizer such as chicken manure pellets or seaweed extract.

Gardening by the moon

Another interesting method of gardening that is related to organic gardening, and that I have dabbled in, is growing by the moon. (There is a connection with bees, so please bear with me!) It has been suggested that the moon phases influence the way in which plants grow. There is circumstantial and some empirical evidence to support this, but probably not enough for the modern scientific mind, which is why, I suspect, it is still regarded as a little 'quirky' and has not become a broadly-acknowledged system of growing. It does have a long pedigree, however. Hesiod, writing in the ninth century BC, tells us that: 'the thirteenth of the waxing month is a bad day to start seeding', and in 1615, Gervase Markham instructs the 'English Hous-wife' [*sic*] that: 'in the new of the Moone shee may sow Garlicke … Marigolds and Time. The Moone full shee may sow … Fennell and Parslie' [*sic*].

In a nutshell, it is posited that, depending on the phase of the moon and other planetary positions, there are optimum days during each month for planting and harvesting plants. My own back garden experiments do bear out some of this hypothesis. Of course, this cannot be regarded as evidence but I have had some bumper crops of potatoes and basketsful of peas using these methods. In addition it appears that plant metabolism and the water content of plants is greater at the full moon, whereas the concentration of carbohydrates is higher at the new moon phase. Additionally, research has shown that bees are more active at around the time of the new moon. I cannot help but think that the imagination does not have to be stretched too far to establish a link between the higher level of plant carbohydrate (including nectar), lower level of plant water content, and the increased activity of bees at new moon.

Whatever system of growing you choose, though, there are some specific jobs that we all need to get stuck into at some stage, so let's have a look at a few of them.

Ground preparation

As with many jobs, preparation is the key to gardening. The most important thing to do if you are planting up a new area or replanting an old one, is to make sure that the spot is completely free of weeds, both annual and perennial. The annual ones are easy to deal with; all you need to do is to hoe them off and leave them to dry in the sun (I should be so lucky, living in Lancashire!) or gather them up and put them on the compost heap; you could also just dig them up, but under no circumstances allow them to set seed.

Perennial weeds are a different matter. If they have a foothold (or should it be roothold?) in your garden then there are ways to be rid of them (unless they are *Equisetum arvense* (horsetail) which will defy any attempt at eradication. This is not surprising, since the root system can penetrate down to 30 feet and it has survived as a genus since the Carboniferous era – approximately 360 to 300 million years ago). If the weeds are pernicious – and your conscience allows – you may have to use a systemic herbicide. A systemic herbicide is a weedkiller that you spray or paint on the leaves of the weed (making sure you follow the instructions on the container to the letter). The chemicals are absorbed by the plant and are taken down to the roots so the entire plant is killed off, not just the parts that are in direct contact with the killer. It does the job but you have to be careful that you do not accidentally get weedkiller on plants close by that you do not want to get rid of.

The chances are that if, like me, you are the sort of person who wants to create a habitat suitable for beneficial insects, you baulk at the idea of anything (particularly chemicals) that kills indiscriminately. In this case there are two ways of going about ridding yourself of perennial weeds. First you can dig them up. But you must get every last piece of root out, otherwise, like the Terminator, *they will be back*! Second, you can cover the entire area with something that will keep out the light (like black plastic or thick cardboard) – and wait. The idea is that if you deny plants one of the essential things they need to grow – in this case, sunlight – they will eventually die. This method takes a bit of forward planning but in my experience it is more or less foolproof. Any plants that do still survive are very much weaker and easier to dig out.

Once you have a clean bed (sounds as if I've changed the sheets!) you can turn over the soil, digging in some organic material like compost or manure, and breaking up any large clods as you go. Firm it down, rake it over and you're ready to plant. Sounds easy, but it can be back-breaking work, especially if you have heavy soil. However, there is a short cut – and a medium cut.

The medium cut first: you can just do the first stage – the digging bit – in the autumn, leaving the soil in large clods. Leave it over the winter and by spring the winter weather, and in particular regular bouts of freezing and thawing, will have broken down the lumps. With a quick rake over, your bed will be ready – that's the theory, anyway.

The short cut is even easier. It is still advisable to get rid of the perennial weeds first, but I have used this method without having cleared the bed and had surprisingly good results. Spread the area with compost or manure to a depth of about 5cm and cover the entire bed with black plastic, making sure you secure the edges with weights (bricks are perfect for the job). Leave it for a minimum of six weeks, or over winter if you start off in the autumn. At the end of the period, the worms will have taken the manure down into the heart of the soil, 'tilling' it as they do so, leaving you with a lovely, rakeable tilth once you take the covering off. You are now ready to plant up.

Conditions in your garden

Choosing bee-friendly plants for your garden is only part of the choice you have to make. You must be aware that not all plants require the same growing conditions. Planting the right plant in the wrong place could prove both disappointing and expensive! It is worth looking carefully at the area you want to plant before you buy anything.

Is your garden in sun or shade?

The majority of bee plants grow best in full sun, and bees often neglect flowers growing in the shade even though they are ones which we know are attractive to them. In other words, choose a sunny spot if you can.

Is your garden exposed or sheltered?

More plants will thrive in a sheltered area than in one exposed to the elements. This is especially true if you live near the coast, which brings with it a whole new set of challenges. It is not usually the temperature that is a problem on the coast, but strong, salt-laden winds, which can play havoc with plants. For your plants to stand any chance of being able to thrive rather than just survive, it is best to provide some sort of shelter in the form of a sturdy hedge, or a man-made construction like a fence. Two of the best shrubs for coastal conditions are *Elaeagnus pungens* 'Maculata' and *Euonymus japonicus*: both are evergreen and as tough as old boots. They will not only make a good hedge but will also make a decent backdrop for any planting in front of them.

Although these shrubs are tough customers it is worth giving them a bit of a helping hand to get established by erecting a temporary permeable screen on the windward side. Ram some posts into the ground, about 1.8m apart and attach some plastic windbreak to it. (This is readily available via the Internet if you can't find any locally.) It might look a little unsightly to begin with but it will help no end in getting your permanent, living windbreak growing.

If you decide to put up a fence, make sure it has a broken surface. By that I don't mean that it is about to fall down, but that it has gaps in it so that the wind can filter through it. The strength of the wind can be reduced by up to 50% by putting a filtering mechanism in its way. If you have a fence with a solid surface, the wind will just hit it and be channelled over the top, forming an eddy on the other side which can cause as much damage to plants as leaving them directly exposed to the wind.

What is the soil like?

Is the soil in your garden free-draining or consistently damp? We all dream of having soil that is free-draining and moisture-retentive at the same time. We want our soil to be able to hold just enough water to satisfy the needs of our plants without being so wet that their feet are constantly soggy. This is actually achievable, but often only after a lot of hard work by the gardener, digging in a good deal of organic matter.

Also, do you have neutral, acidic or calcareous soil? To get a first impression of what your soil might be, have a look at what plants thrive in your neighbours' gardens. If you are surrounded by camellias and rhododendrons then the soil will be acidic and there will be no point in trying to grow pinks or other lime-loving plants. It is worth investing in an inexpensive testing kit from your local garden centre which will give you a reasonably accurate indication of what pH your soil is. (A pH ['potential of Hydrogen'] measurement will tell you how acidic or alkaline the material is. To give you an idea, water measures roughly 7 and is neutral; vinegar is 2/3 and is therefore acid; ammonia is 11/12 and is alkaline. Soils in Britain usually fall between pH 4 and 8.5; most plants prefer a neutral soil of pH 6.5 to 7, which is the range in which nutrients are most easily available.)

Growing in containers

One way to avoid being restricted by the type of soil you have is to grow plants in containers. This way you have control over the growing medium, which means that you can grow whatever you choose. In addition, containers can be moved around, so as well as being 'mini gardens' in their own right they can also be useful in a much bigger garden to fill in the gaps. I usually have several pots of spring bulbs or summer-flowering lilies ready to bring a fillip of colour to a drab space at the end of my garden where little else will grow.

There are three main disadvantages of growing in containers, however. The first is the lack of physical space, which will preclude growing any very large or deep-rooted plants. Second, the plants will soon use up the available nutrients in the compost and because their roots are restricted they cannot search down into the soil for food. You will have to provide food for them in the shape of watered-down comfrey fertilizer, slow-release fertilizer pellets or another proprietary plant food. In the latter two cases, follow the recommended application rate on the packet – too much food is almost worse than not enough. The third disadvantage is watering. Moisture will evaporate far quicker from a container than from open soil and during the summer months your pots will undoubtedly have to be watered every day and sometimes twice a day. As a rough guide, 2·5cm (1in) of water will penetrate about 15–20cm (6–8in) of soil so you can see that even if it rains you will still have to top up your containers.

Planting trees and shrubs

As long as the ground is not frozen and is not too wet, bare-root trees and shrubs, especially roses, can be planted during the dormant period which is usually from November to March. Bare-root plants are ones that have been grown in a field and then dug up. They are usually much cheaper than plants grown in pots because they are less labour-intensive to grow, and are easier and lighter to move around, therefore reducing transport costs. Most bare-root plants are available at specialist nurseries, by mail order or over the Internet. Garden centres rarely stock them.

The roots are usually trimmed and washed clean. It is vital to soak the roots of your plant in water for a couple of hours as soon as you receive them and then get them into the ground, either in their permanent position, or, if this is not possible, you must provide them with a temporary home by 'heeling them in' somewhere. This involves digging a shallow trench big enough to be able to cover the roots of the plant with loose, friable soil, sand or compost. Your plants will be quite happy for as long as is necessary: it is better to wait until the conditions are good to plant your tree or shrub in its permanent position rather than try and do it in adverse conditions when your 'baby' will get a poor start to its new life in your garden. If you are able to plant it on the same day as you receive it then so much the better.

Planting is straightforward. Dig a hole slightly larger than the root system and deep enough so that the plant will be at the same depth as it was growing in the nursery and fork over the bottom. (Do not plant trees too deeply otherwise they will rot – look for the 'water line' on the trunk and make sure the soil is at the same level. Roses, however, should be planted 'deep'; the base of the stems should be about 7.5cm below soil level.)

If you are planting a tree the chances are it will need staking: put the stake in the hole before planting to avoid damaging the roots. Place the plant in the hole and back-fill with soil, enriched with a little fertilizer (I use pelleted chicken manure), making sure that you firm the soil around the roots as you go along. Then water, even if the soil is moist.

You can plant containerized trees and shrubs, and indeed, herbaceous perennials, at any time during the year as long as the ground conditions are suitable, namely that it is not frozen or too wet. Containerized plants require the same conditions as bare-root material but they will not need to be 'heeled in' if the conditions are bad – just keep them in their pots until conditions are good enough to plant them.

Watering

Other than keeping my containers moist I use water sparingly in the garden. I do give new plants a generous watering-in and keep them modestly watered during the first season until they are established but thereafter they have to fend for themselves. The exception is some vegetables (such as tomatoes and courgettes) which do require water in order to produce decent crops. My garden is relatively small so all my watering needs are catered for from the water butts that collect rainwater from the roof of the house and garage. I have three 'in tandem' collecting water from the house so that when the first one is full, it overflows into the second one and so on. Only on rare occasions have I had to draw on mains water but that might have something to do with the fact that I live in Lancashire where it does rain quite a bit!

Tidying up the garden

Jobs that must be done wherever you live are clearing away debris and improving the soil. Many gardeners do their tidying up in autumn: clearing away leaves, cutting back spent flowers, and mulching. I do clear away leaves during the autumn and early winter for two reasons: firstly because if you leave them on the ground they make a cosy cover under which slugs will hide; and, secondly, because I like to collect the leaves together to make leaf mould. (Leaf mould is just rotted down leaves. It is not a quick process, but well worth the wait.) The other two jobs – cutting back flowers and mulching – I tend to do in the spring. Unless the spent flowers are particularly unsightly, I leave them where they are, mainly because they provide overwintering sites for insects, but also because some of the flower heads are quite attractive (I have heard them described as 'dying well'), especially if we have a cold spell and they are covered in hoar frost.

Mulching

I nearly always leave the mulching until spring as I do not often get round to it before the colder weather sets in. Mulching in the spring means that the winter moisture can be trapped in the ground just as it is warming up. I like to mulch a little bit earlier actually, when the spring bulbs are just beginning to show. They are quite happy to have another couple of inches of material to grow through and if it were left any later they would be far too advanced to be able to be mulched around.

Weeding

Well, as a gardener you either love it or hate it. I would rather not have to do it but recognize that if I don't I shall be in trouble later in the season. I always think that I have better things to do with my time – like looking after my bees! – than hoeing off or digging up plants which happen to have chosen a place to grow in that doesn't meet with my approval. After all, as has been said many times before, one definition of a weed is just a plant growing in the wrong place. There is a way to mitigate the weed invasion, however, and that is to have mulched earlier in the year. Even if overwintering weed seeds lurking underneath the mulch have germinated, there will be too much material for their skinny shoots to push through and they will perish before they reach the surface. Hopefully, you will have eradicated perennial weeds before having planted up. If the odd one has managed to go undetected, then the best way to deal with it is just to dig it up.

Pest control

Other than weeds, the other group of 'baddies' in the garden is pests. I do not particularly like them but I am inclined to take the view that if the ecology in your garden is balanced then there will be no need to use bug killers anyway. So far this principle has worked for me but I am expecting a plague of locusts of biblical proportions any day now!

Increasing plant numbers

There is real enjoyment in wandering around a well-stocked nursery. My garden is full to overflowing with beautiful, bee-friendly plants but I know that I can squeeze in just one more if I really try. (A bit like making room for that last, lonely chocolate eclair that's left on the cake plate at tea time …) And then you get home with your new treasure and it sits in its pot for a few days while you stroll around the garden trying to find a suitable place: there, next to the rosemary – no, too dry; how about by the garage – no, too shady. Until you give up and it finds its way into the crate with all your home-raised plants waiting to go to the local charity plant sale.

Cuttings

I am an inveterate cutting-taker. I do not mean that I go around other people's gardens pinching off – and pinching – suitable material to use for cuttings; certainly not. But I do take lots of cuttings from my own garden to grow on and give to friends or to donate to charitable events. There is nothing more satisfying than being able to share some of your best-loved specimens with others, and taking cuttings is one of the easiest ways of doing this. There are a number of types of cuttings: softwood, basal, semi-ripe, hardwood and root.

A word of warning, however: do not expect all your cuttings to 'take'; namely, to produce new roots and shoots. Some will inevitably die before they have a chance to live, as it were, so don't be disappointed if you lose a good number. As a general rule of thumb, I always take twice the number of cuttings as the number of plants I would like to have – I usually end up with a better success rate than that, but I like to be on the safe side.

Arguably, you can increase the success rate by using a hormone rooting gel, liquid or powder which contains plant hormones and usually a fungicide which, respectively, causes stem cells to grow and develop, and reduces the chance of fungal infection. I say arguably because I didn't seem to have any better results when I used a rooting gel than when I did not, although I did use one approved for organic growing, which might have made a difference. Have a go yourself and see what happens.

Softwood cuttings

In spring you can take softwood cuttings from shrubs. Material is taken from the soft and flexible young shoot tips, which root readily at this time of year. The best time of day to do this is in the morning when the plant is turgid, or full of water. Before you start taking cuttings, make sure you have a clean, sharp knife and have ready some clean, 9cm pots filled with compost. The number of pots depends on how many cuttings you wish to take: you can fit four or five in a pot. Select a non-flowering shoot and cut it just above where the leaves join the stem – the node. You need a shoot about 10cm long. Pop the shoots into a plastic bag until you are ready to prepare them for planting: this will prevent them from wilting. When you are ready to plant them, take them out of the plastic bag and, with a sharp knife, cut the bottom of the shoot off to just below a node. Carefully remove the lower leaves and pinch out the top of the shoot. Then make a hole in the pot of compost near to the edge – you can use a dibber, but I use a pencil, which seems to do just as good a job – pop the cutting in so that the first set of leaves is above the surface of the compost and push the compost back around the cutting. Water the pot from above so that the compost settles around the cutting.

If you have a heated propagator this is ideal, but I pop each pot into a large plastic freezer bag and use the thin paper-coated wire tie to loosely pull the opening together at the top. I do not tie it tightly, but leave a small gap so that a certain amount of air can still circulate. The whole thing then goes on the windowsill and, because the pot is inside the bag, you don't need to worry about standing the pot on a saucer. Do not put the pot in direct sunlight, though, as this will scorch the leaves. Open up the bag periodically to aid ventilation and remove any dead or decaying material as you see it. Keep the compost moist. The cuttings should develop roots in about ten weeks and once they are well rooted you can pot them on into their own individual pots. As long as you can do this by midsummer they will develop enough roots to carry them through the winter; otherwise leave them and pot them up next spring.

If you want to take cuttings from tender perennials, such as *Penstemon* (page 36) and *Verbena rigida* (page 59), follow the same procedure for taking softwood cuttings, but make sure you keep them undercover in a light, frost-free environment at all stages, until the following year when they can be planted out in the garden when there is no danger of frost.

Basal cuttings

A similar method to taking softwood cuttings is used to take cuttings from some perennials. The only difference is that the cuttings are taken from the base of the plant where the young shoots start to grow in spring: these are called basal cuttings. Take the shoots from as close to the base as possible and then treat them just as you would softwood cuttings.

Semi-ripe cuttings

Late summer into autumn is the time to take semi-ripe cuttings. These are cuttings from the current season's growth where the base of the cutting is hard while the tip is still soft. The method for taking semi-ripe cuttings is just the same as for softwood cuttings, above.

Hardwood cuttings

It is important to remember to take hardwood cuttings during the dormant season, when the plant is not growing. The ideal time is just after the leaves have fallen in the autumn or during late winter to early spring, just before the new buds start to burst. To take hardwood cuttings you should choose a shoot that has grown in the current year. If you select a long enough one, you may be able to get more than one cutting from it. Remove the soft growth at the tip and cut the rest into sections 15–30cm long, depending on how many dormant buds there are. (Both roots and shoots will break from the dormant buds so the more buds there are, the shorter the cutting can be.) You should make a sloping cut just above a bud at the top and a straight cut just below a bud at the base – this way you know which way up you should plant the cutting.

You can plant the cuttings into a trench outdoors but I always put mine into pots filled with a mixture of 50% horticultural grit and 50% multi-purpose compost. Water them in and leave them in a sheltered position in the garden (a cold frame is ideal but not a necessity). Do not worry if nothing appears to be happening for some time after you have planted your cuttings; it usually takes a while for them to develop roots and shoots and it may be well into the new growing season before anything starts to occur.

Root cuttings

You can take root cuttings from mid-autumn to early winter. First, fill a seed tray with gritty compost, press it down slightly and water it. Then carefully dig up the parent plant, keeping intact as much root as possible. With a sharp knife cut off vigorous roots as close to the crown as you can, but do not remove more than a third of the root system from the parent plant otherwise it will struggle when you replant it. Cut each root into 3–10cm lengths (the

Verbena rigida

thinner the root, the longer the cutting should be), place them on the surface of a prepared tray about 4cm apart, and cover them with a thin layer of compost. Put them in a sheltered position – again a cold frame is ideal – and in the following spring, when there are signs of growth and the cuttings are well rooted, pot them up individually to grow on.

A friend of mine has adopted another way of taking root cuttings which has proved as successful as any other she has tried. She prepares the cuttings in the usual way, but rather than putting them in seed trays, she simply puts them in a plastic bag with several handfuls of only-just-damp compost. She then gathers the top of the bag together, blows in it so that it is full of air, and seals it with a wire tie. She leaves it in a cool place until she sees shoots and roots appearing, and then pots up the little plantlets. Easy – and, she maintains, foolproof!

Division

Another method of increasing your plant numbers is by division. This simply means taking an existing plant and splitting it into pieces.

The process is straightforward: carefully dig up the plant, trying not to damage too many roots, and shake off as much soil as possible. Then all you have to do is to divide the plant into sections, making sure that each section has three to five healthy shoots. Sometimes you can tease the clumps apart with your hands. Some plants, however, require something a little more vigorous and you may need to use a sharp knife or even a spade to slice the plant into sections. It will become clear as you do it which technique you will need to use. Once you have your new clumps you can plant them straight away, making sure you water them in well. If you do not want to plant them in the ground immediately, you can pot them up individually and keep them until required – again make sure they are well watered.

Sowing seeds

Another easy, and very cheap, way to increase the number of plants you have is to sow some seed. When it comes to annuals, this is really the only method to raise new plants because, by their definition, they are ones which germinate, grow, flower, and set seed in one season. You can raise other plants, such as perennials and even shrubs and trees, from seed but these take much longer to reach maturity. Seed sowing can be divided into two main sections: sowing seeds indoors and sowing seeds outdoors. When you buy seeds, check on the packet as to which method is best.

Sowing seeds indoors

Indoor sowing is suitable for a number of plants, particularly tender ones which can be started off and grown on inside so that they have a 'head start' when they are planted out in the garden when the weather is warm enough. Many vegetables benefit from indoor sowing, too, such as runner beans, tomatoes and courgettes. You can start sowing these in February or March.

The container you use will depend on the size of the seed. Small seeds can be sown into shallow trays and then when they are big enough they can be transplanted into bigger pots. Larger seeds can be sown into trays with individual modules; this causes less disruption to the root system when you replant them. Very large seeds can have their own individual 9cm diameter pots right from the word go.

You can mix your own seed-sowing compost but I find that a proprietary compost which is specially formulated for seeds is the best: it is sterile and is not nutrient-rich. Fill your container with compost and gently firm it down. It is best to water at this stage: if you water after you have sown the seed, the water may wash the seed together, especially if it is fine. Small seeds can be scattered across the surface of the compost; larger seeds can be pushed into the surface. I then like to cover the seeds with perlite or vermiculite, rather than compost. These provide sufficient coverage without staying too moist.

Now all you need to do is cover the container with clear polythene, a sheet of glass, or its own plastic cover, and put it in a warm place. You do not need an elaborate greenhouse for this; anywhere with a fairly constant temperature and out of draughts will suffice. Much to my husband's chagrin I use my kitchen windowsill (he's a very tidy person; I am not): you can keep a constant eye on your 'babies' and you would be surprised how many small seed trays you can fit on it!

Once the seeds have germinated you can remove the cover so that the seedlings have good ventilation. When they have their first 'true' leaves you can pot them on into individual modules or 9cm pots respectively, depending on whether they started off in trays or modules. You can use multi-purpose compost for this task, although I like to supplement it by putting a chicken manure pellet near the bottom of the module or pot to give the plant a little more in the way of nutrients as it grows. As soon as the plants are big enough, and the risk of frosts has disappeared, you can start hardening them off. This means gradually getting your babies used to the outside world. The easiest way to do this is to begin by putting them outside in a sheltered spot during the day and then bringing them back under cover at night. As the plants begin to 'toughen up' you can leave them outside all the time, but you must still give them some protection at night by covering them with some horticultural fleece. The hardening off process should take between two to three weeks, depending on the weather, by which time pesky frosts will be but a memory and the plants will be strong enough to be planted out in the garden.

Sowing seeds outdoors

Many seeds are best sown straight into the soil. Some actually prefer it because they do not like their roots to be disturbed, but check the seed packet to see what is recommended. You can sow seeds outdoors as long as the soil is warm and moist, and the air temperature is not too high, which in practical terms means either mid-spring to early summer, or late summer to early autumn.

It is vital to have a good seedbed, free of weeds and with a fine tilth (think of the trays that you prepared for sowing indoors). With the corner of a hoe or rake make a drill, or shallow depression, in the soil, the depth of which will depend on the seeds you are sowing. Look at the seed packet for this information. There you will also find out how far apart you should sow the seeds and space the rows. Then water the drill. Scatter or place the seeds evenly along the drill and then cover them with a thin layer of soil.

You will need to check occasionally if the seeds need any water. If you have sown the seeds too thickly, you may need to thin them out when they are big enough. This simply means removing seedlings so that you are left with single plants, evenly spaced, with enough room between each for them to grow on.

Labelling

Whenever you take cuttings, divide plants, sow seeds – in fact anything that involves putting something into a pot or tray – make sure you LABEL IT. This is of such vital importance that it deserves to be in capital letters. No matter how good your memory is, I guarantee that you will have no idea what you put in which pot in even three weeks' time, never mind three months. If you entrust this task to a helper it is always a good idea to make sure that *they* know exactly what it is too.

I recall the tale of a young lad (an intelligent boy who was eager to please) who wanted to help his grandfather sow some seeds. Grandfather's eyesight was not as good as it used to be and although he could see well enough without his spectacles to distinguish the picture on the packet, scatter the seeds in the tray and cover them over, writing the labels was a challenge. On this occasion he had mislaid his specs so he asked Frank (not his real name) to do the honours, which Frank was only too willing to do. 'What shall I write on the label, Gramps?' asked Frank, with pencil poised. 'Well, Frank, my boy, these are *Mesembryanthemum criniflorum*,' said Gramps. As I mentioned, Frank was a clever boy but, being only eight years old, he struggled with pronouncing *Mesembryanthemum criniflorum*, let alone spelling it. He did not want to lose face in front of his grandfather, whom he idolized, so he pondered for a moment or two what to do. Then it came to him: he wrote 'daisy' on the label which, actually, was not so far off the mark since *Mesembryanthemum* is the Latin name for Livingstone daisy. It had Gramps scratching his head some weeks later, though, when he came to pot on his seedlings.

Urban gardening

I said earlier that gardening is pretty much the same wherever you live, but I would like to say a word or two about gardening in an urban environment. It has much in common with gardening in the suburbs or in the country but there are two main differences, namely climate and space.

Climate

In a town or city, open spaces – no matter how large or small, whether public or private – are enclosed. This may seem contradictory, but invariably a garden, backyard or park will be surrounded by a boundary, usually in the form of buildings, walls, fences or hedges. This means that not only will the enclosed space be protected from the worst of the weather, but also the temperature is often 2–3°C (and can be as much as 6°C) higher than elsewhere. These factors provide a microclimate where the urban growing season can often be warmer and longer than in rural situations. This will have an impact on what plants will thrive and also on the length of the 'bee' season.

Space

There are two types of space within an urban environment: public and private. Ironically, there is usually more public space in an average town or city than in a corresponding measured area of countryside. It is true that large parts of the countryside are accessible to the public but they are usually privately owned. In towns, however, parks, pedestrian areas and even roundabouts are municipal, owned and maintained by the local council or other civic body. If these areas are landscaped then decisions on planting are made by the relevant amenity department. Often the planting is designed to be low maintenance (lots of shrubs) or very colourful (annual bedding) with, arguably, little thought being given to wildlife – and especially bee support.

I visited a flower show recently and came across some bedding displays by council parks departments. A huge amount of effort and skill had gone into creating these exhibits and I was interested to find a beautiful display of annuals, laid out to depict hives and honeycomb, highlighting the plight of bees. It was a splendid concept and design – the problem was (and it was a big problem to my mind) only one of the plants was bee-friendly. In fact the greater part of the display comprised double *Tagetes* (African marigolds) which, although they were ideal specimens to represent the golden honeycomb, were inappropriate as far as providing food for bees was concerned. I could not help but think this particular council had come unstuck a little with this exhibit.

As we shall see later, there are numerous shrubs and annuals that meet all the criteria for bee-friendliness. If, therefore, you find that your 'public' planting is not bee-friendly then you can petition your council to make it so. The vast majority of public areas in towns and cities are very well maintained and the amenity departments do a remarkable job.

Private space in towns and cities is often limited although a large number of houses do have substantial gardens, both front and back. Other residences may have smaller gardens, balconies, space for window boxes, or may even accommodate a roof garden. Whether you have a large back garden or a simple window box you can still provide forage for bees, which in itself may not sustain a colony. However, if everyone in your street, and everyone in every street in your town, earmarked a small area to grow some bee-friendly plants then the total area would amount to a considerable resource for bees.

Gardening in the country

Living out of town often means that you will have a larger garden in which to indulge your passion for growing things. You have space for that pristine potager; you can have a separate 'spring' garden; and just look at the depth of those herbaceous borders! But before your daydreams carry you away on a flight of fancy, let me bring you back down to earth by telling you that living in the countryside does not mean that your gardening life is all plain sailing. There are a whole raft of inconveniences and disasters waiting to shipwreck the unwary gardener. Flopsy, Mopsy, Cottontail and their big brother Peter Rabbit lurk behind the shed, waiting until you pop indoors to make a cup of tea and then, crunch – there goes another row of carrots. Someone leaves the gate to the chicken run open and not only have the hens eaten up all the slugs (hooray!) but what was once a precious bed with emerging flower seeds now looks like a barren wasteland where the hens have scratched and used it as a dust bath (boo!).

One of the most heartbreaking things that I have witnessed, though, was when one of my neighbour's cows (I once lived in an old cottage next to a farm and, believe it or not, the farmer's name was actually MacDonald, but he was not very old) decided to breach my combined fence and hedge defence to meander along my herbaceous border, sampling a few plants as she did so. I had spent the best part of three years working on that border until I thought I had something that even the late, great Geoff Hamilton would nod in approval at, and it took Cowslip only a little more than three minutes to destroy the lot. Cloven hooves stomping on clove pinks are to be avoided.

Even worse than Cowslip, though, is the weather. In the countryside temperatures can fluctuate dramatically and wind can buffet precious plants. Sometimes, even despite your best efforts to mitigate calamity, disaster strikes in the form of an unforecast frost or unexpected gales and your garden succumbs. All you can do then is to survey the damage, take stock, and try to bring some sort of order to the chaos.

Gentle chaos

Talking of chaos, I would like to make a general plea for *gentle* chaos in your garden (to paraphrase Mirabel Osler who has written an engaging book entitled *A Gentle Plea for Chaos* – well worth a read). I do not advocate letting your garden go, allowing nature to take over – even apparently 'wild' and 'natural' landscapes are carefully managed – but I would encourage you to be less fastidious about certain aspects of garden maintenance. The lawn, if you have one, for example. No matter how many times you weed and feed, scarify and roll, cut and rake, the average garden lawn will never reach the exacting 'bowling green' standards too many of us try to make it achieve. Why not set the standard a little lower and the cutter blade a fraction higher to allow a few daisies, clovers and – dare I say it – dandelions to flower? As we know, *Taraxacum officinale* (dandelion) are high on the list of bee-friendly flowers.

Why not also tolerate a few wild flowers in the perennial border? *Knautia arvensis* (field scabious) and *Salvia pratensis* (meadow clary) are every bit as good as some cultivated varieties and if you do not want them to seed themselves everywhere then by all means deadhead before the seed ripens.

If you have a herb patch, or grow herbs in pots or in a window box, allow some of them to flower. You may lose some of the delicate pungency of the leaves but it will be worth the sacrifice because an amazing number of herb flowers are superlative bee plants. I am thinking of *Origanum vulgare* (marjoram) (page 25), *Hyssopus officinalis* (hyssop) (page 67), *Mentha* (mint) (page 68), *Salvia officinalis* (sage) and *Thymus* (thyme) in particular.

And then there's traditional summer bedding: regimented, straight rows; vivid, contrasting colours. You either love it or hate it. Bees don't mind straight rows, but they do care about what flowers are in those serried ranks – have a look at the gazetteer for some suggestions for annual plants. There is no reason why you can't have straight rows, vivid, contrasting colours, and still be bee-friendly.

3
Plants and Gardening Jobs
Season by Season

The life of a gardener, like that of a bee-keeper, is determined by the rhythm of the seasons and the inexorable pulse of spring following winter, autumn following summer. But the boundaries between each season are not definite: the margins are blurred and just because we theoretically enter autumn during September, it does not mean that all the summer flowers will curl up and die. There is bound to be some overlap, especially if the weather decides to do strange things, like temperatures of 20°C plus in April and then monsoon-like weather in July. So please treat the suggestions in each section as just those: suggestions rather than instructions. Let's now look at each of the seasons and see what flowers are available, and what jobs can be done in the garden. We could start with any season but spring is traditionally the time of new beginnings, so let's start there.

Spring – March to May

Spring can creep up on you. It sometimes feels that although it was winter last week it is now spring. There is a Chinese proverb that says: 'Spring is sooner recognized by plants than by men.' It is true. Long-dormant plants seem to sense before we do that the nights are getting shorter and the days warmer, and now is the time to start thrusting new shoots through the soil and letting buds burst from their casings. Bulbs start to colour the ground and to fill the air; fruit trees are clothed with blossom far showier than any designer outfit. And bees begin to emerge from the hive – slowly at first, with scout bees being sent out to reconnoitre and report back if they find anything of worth.

It is at this time of year that, if you're in the northern hemisphere and you happen to study the night sky, you will see the Beehive cluster reappear in the constellation of Cancer. I like to think that the timing of the appearance of both the heavenly Beehive and the terrestrial bees is not merely a coincidence, but perhaps this is just fanciful.

Spring plants for bees

Early spring can be a difficult time, especially if we have had a severe winter. Precious food sources for the bees are sometimes reluctant to make an appearance. Some of the first flowers to emerge in March and through to May are often those which humans pass by without a second glance, but are a beacon for bees. These are the flowers of *Salix* (willow, especially *S. caprea* for a large garden or *S. lanata* for a small garden) which are an important source of pollen and can also provide modest amounts of nectar. Be sure to buy a male plant, however, as the female will bear no pollen.

Flowers that are hard to miss by both humans and bees are those of the *Crocus* (particularly *C. chrysanthus* and *C. tommasinianus*). These give some nectar and are an excellent source of pollen, which bees need at this time of year to build themselves up – a bit like a tonic. Other spring-flowering bulbs and tubers will also provide a little food for bees, notably *Galanthus nivalis* (snowdrops) (which are best transplanted immediately after flowering when they still have their foliage, known as 'in the green'), *Muscari* (grape hyacinth) and *Eranthis hyemalis* (winter aconite).

T.S. Eliot reminds us that 'April is the cruellest month'; from a gardening point of view it sometimes appears so. Whatever the weather we feel hard done by: if we have too much rain we cannot sow seeds; if we don't have enough the plants will not grow. It can be cruel to the bees, too. Many early-flowering plants have already had their day and if the weather is adverse then late spring flowers are unwilling to emerge. Nevertheless, *Mahonia aquifolium* (Oregon grape) can always be relied upon to fill the pollen gap with vivid yellow flowers scenting the air between February and well into May. Before long, however, nature puts on one of her most attractive and alluring shows – blossom time.

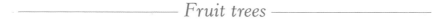
Fruit trees

Prunus domestica (plums) and *Prunus domestica* subsp. *insititia* (damsons, also known as just *Prunus insititia*) are the first to show in March or April, followed by *Pyrus communis* (pear) and *Prunus avium* (cherry). You will see *Malus* blossom (apple) throughout April and into May, and *Cydonia oblonga* (quince) will wait until late April before it flowers. Bringing up the rear is *Mespilus germanica* (medlar) (page 197) in mid May.

A friend gave me a haiku poem recently which I think encapsulates the vital relationship between fruit trees and the bee:

> *Fruit blossom unfurls,*
> *Golden honeybee alights:*
> *The precious union*

It is without doubt a precious union; without bees to pollinate the flowers we will have no fruit.

Orchards of top-fruit trees in full bloom are certainly a sight to behold, and in the Vale of Evesham in Worcestershire you can follow the Spring Blossom Trail through the countryside, taking in a real confection of blossom. But you don't need to live in the country or have a large garden to have a fruit tree: even a modest back garden in town or country can be home to a dwarf tree. The vigour and eventual size of the fruit tree is entirely dependent on the rootstock on to which the top, fruiting part has been grafted.

Some fruit trees, in particular apple and pear, can also be trained into different forms. I particularly like the espalier form (*espalier* is a French term, derived from the Italian word *spalliera*, meaning to rest against a shoulder). Espalier-trained trees consist of a set of horizontal 'arms' extending out either way from a main, vertical stem. The 'arms' have short lateral branches or spurs on which fruit is produced. The tree is trained into shape and held there by growing it against a wall or fence, or, if you want to use it to divide part of the garden, it can be trained against a trellis or wires between posts (page 73). They are very productive and decorative at the same time. You will also find 'step-over' trees: these are simply espalier trees with only one 'arm' low down, on either side of the main stem.

Other trees

Other trees are also flowering now but these are often overlooked as valuable sources of bee food. *Aesculus hippocastanum* (horse chestnut), *Crataegus monogyna* (hawthorn), *Sorbus aucuparia* (rowan) and *Acer pseudo-platanus* (sycamore) all provide pollen and varying amounts of nectar. These take up a lot of space, however (up to 24m!), so unless you have a large garden, they are best admired in the surrounding landscape.

Spring shrubs

There are other plants whose flowers are available to bees during spring. Shrubs, such as *Chaenomeles* (ornamental quince – not to be confused with the edible fruit tree), sweetly scented *Daphne* (*D. mezereum*) and early *Viburnum* (*V. tinus* (page 73) and *V.* x *bodnantense* in particular) provide pollen and nectar in varying degrees. Early flowering *Cytisus* (broom), *Ribes sanguineum* (flowering currant) and *Rosmarinus* (rosemary) also help to break the bees' fast as do some early climbers such as *Clematis* (*C. armandii* and *C. montana*). Of all the shrubs in flower at this time of year, however, *Ulex* (gorse) is the one that will be most attractive to bees.

Spring bulbs

Other than the flowering bulbs that I mentioned earlier, mid to late spring will find *Tulipa* (tulip) (pages 74–5) taking centre stage in many gardens. Although not a top-ranking bee flower, tulips do provide masses of pollen which bees will search out if there is nothing better on the menu.

Perennials

You can also find some early perennials showing their faces in spring: *Lamium maculatum* (ornamental dead nettle), *Polemonium caeruleum* (Jacob's ladder) (page 76), *Doronicum* (leopard's bane) and some varieties of the hardy *Geranium* (page 76). One useful perennial which is invariably grown as a biennial (a plant which in effect grows in the first year and flowers and dies in the second) is *Cheiranthus cheiri* (wallflower) (page 77). It was once a stalwart of spring bedding, giving a beautiful, scented display – and lots of nectar and pollen – during April and May. It seems to have been superseded by pansies and polyanthus nowadays since they can be relied upon to give colour much earlier and for longer than the wallflower. But our gain is the bees' loss because they yield next to nothing in terms of bee food.

Apple tree trained as an espalier
Viburnum tinus

Tulipa sp. (tuplips) – *previous page*
Polemonium caeruleum (jacob's ladder) – *above*
Geranium sp. – *below*
Cheiranthus cheiri (wallflower) – *opposite*

Wild flowers

There are, however, some wild flowers blooming in spring that are invaluable to bees. One of the first to flower is *Tussilago farfara* (coltsfoot) which looks a little like dandelion but it appears without leaves, which sprout in late spring and early summer. Following close behind coltsfoot are *Lamium* (dead nettles) and the ubiquitous *Taraxacum officinale* (dandelion). Later comes *Trifolium repens* (white clover) (page 25), one of the best bee plants because it produces copious nectar over a very long period.

You may also find pockets of *Viola odorata* (sweet violet) being visited by the occasional bee. These delicate, sweetly scented spring flowers are another favourite of mine, mainly because, as children, my sister and I were allowed to go off down Stinky Lane to pick them. We would only gather a few flowers because to get to them was a major feat of daring. We had to jump across what to us was an enormously wide and deep ditch to get to the bank where the nodding violets grew (there may even have been wild thyme). What made it so daring is that the ditch was filled with liquid cow manure which found its way there after the milking stalls at the local farm had been washed down after each session (this would not be allowed nowadays, of course). That's why it was called Stinky Lane. Of course, the best flowers were always those high up on the bank at the point where the ditch was widest. Yes, you've guessed. We both slipped in. Welly boots are fine things when the stuff you want to protect your feet from stays on the outside, but once it reaches, and breaches, the tops it gushes into them like the tide enveloping a sandcastle moat. And I can truly say there is nothing more disgusting than welly boots filled with liquid muck. Needless to say we trudged – or rather, squelched – home *sans* flowers and smelling of something far removed from violets. From then on we left the flowers for the bees.

Spring jobs for bee-friendly gardeners

Bees are not the only creatures up and about in springtime: that perennial of all specimens emerges, bleary-eyed from looking at all those seed catalogues – the gardener. There is much to do in the garden in spring – mulching, planting, sowing seeds, taking cuttings, dividing plants, pruning …

Visit your local nursery

Although all manner of containerized, hardy, woody and herbaceous plants can be planted at any time of year (unless the ground is frozen or there has been heavy rain – mmm, that sounds like a typical winter here in Lancashire!), spring is a good time to visit your local nursery to stock up. Nurseries will have a top-quality collection of plants which have weathered the winter or new-season plants that are raring to go. And don't be tempted to just go for ones that are already in flower, otherwise your garden will look a little forlorn later in the year – do a bit of homework before you go so that you know what you're looking for. If you can't find exactly what you want, be guided by the nursery owner, who will be able to recommend close substitutes. Look for plants with strong new growth and with a root system that fills the pot but is not pot-bound. Knock the plant out of its pot and have a good look – no nursery owner worth their salt will mind you doing this.

Plant trees and shrubs

Both ornamental and fruit trees and shrubs can be planted now. The latter are often overlooked because they are not regarded as 'special', like trees, nor are they as flamboyant as perennials or annuals, but they do provide the 'backbone' of any border or planted area. In recent years there has been a move away from shrubs to ornamental grasses to provide such structure – the so-called New Perennial style. Grasses do have an attractive form, and as exponents of the approach maintain, they do 'die' well, providing year-round visual interest, although in some areas the grasses have swamped the less vigorous perennial flowering plants, losing the 'tapestry' effect of the planting. And, because the grasses are wind-pollinated, they are worthless to bees.

Some of the edible fruit-bearing shrubs are good for bees, though, like the currants and gooseberries. If you do not have room for a separate kitchen garden, then why not consider planting some in the ornamental bed? Although they are deciduous, they will give background structure to the planting, and you will get a crop of fruit to boot! Bees also like other flowering shrubs: evergreen or deciduous, they do not mind, as long as they have pollen- and/or nectar-rich flowers.

Plant perennials

Spring is a good time to plant perennials. Planting them now, when new growth is just emerging, gives them an excellent start and in no time at all they will get their roots down into the ground, and form a goodly mass of leaf and flower buds above ground. It is as well to give tall-growing perennials support before they actually need it. A circle of canes with twine criss-crossed between them, pea sticks, or specially made wire frames will all do a good job, and by the time the flowers have grown through them the supports will be all but invisible.

The 'Chelsea Chop'

Some perennials respond well to what is known as the 'Chelsea Chop'. I was going to say this is not some fashionable haircut, but in a way it is. At about the time of the Chelsea Flower Show (late May) many gardeners cut back the new growth on a number of perennials, such as *Sedum* (page 81), *Rudbeckia* (page 31), *Helenium* (page 81), *Echinacea* (page 82) and *Aster* (page 82). Some merciless, secateurs-toting individuals raze the plants to the ground, leaving the poor crowns looking like skinheads. I cannot bring myself to be so ruthless; I prefer to give mine a short back and sides, leaving some healthy-looking stubble to grow away again. It is worth gritting your teeth and carrying out this procedure, though, because it will result in more flowers on stockier, stronger plants, some of which will not need staking.

Plant tubers

Summer-flowering tubers can be planted in spring, too. I am particularly thinking about *Dahlia* (single-flowered varieties) (page 111), which should be planted in an area which receives full sun. Dig a hole deep enough so that when the tuber is placed in it the 'crown' or top of the tuber is about 8–10cm below the surface of the soil. They should be planted 75–90cm apart. If the variety you're growing is tall, it is a good idea to a put a stake in next to each tuber as you plant them to provide support later. The stakes may look a bit untidy for a while but you'd be surprised at just how quickly the plants will grow, and you may damage the tuber if you try and push a stake in later. Shorter varieties will not need staking. Do not forget to water the tuber in, and (if it does not have a stake to mark the spot) put a label where you have planted it, in case you forget where it is and inadvertently damage the tuber by trying to plant something else there.

Annual bedding

A word of warning: do not plant out annual bedding yet – leave it until much later when all risk of frost has passed. And certainly do not be tempted by special Easter offers at garden centres – even a late Easter is

Sedum spectabile (ice plant)

Helenium 'Moerheim Beauty' (sneezeweed)

Echinacea purpurea (coneflower)
Aster sp. (michaelmas daisy)

still too early to plant out tender annuals. You will only have to buy more when a late frost destroys them – good for the garden centres, but bad for your pocket!

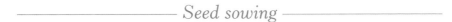
Seed sowing

While we are talking about annuals, some annuals – and biennials and vegetable seeds – can be sown under cover now, although be sure to look at the cultivation notes on the packets of seeds to see which will benefit from such treatment.

Wild flowers

Many wild flower seeds can be sown directly into the soil during late spring. Even a couple of square feet given over to a 'wild' area in your garden will attract any number of beneficial insects as well as bees. I love wild flowers. As a child I lived in a small village in Sussex, where my father had a modest market garden and we were surrounded by pasture which sustained a herd of Jersey and Guernsey dairy cows. Nowadays we would say that the land was farmed organically – then it was just that the farmer was 'old-fashioned'; he 'couldn't be havin'' with all them new-fangled 'erby-sides, pesky-sides and kiss-my-backsides', so his beloved cows ate what nature intended them to eat and they, and the land, were better for it. An abiding memory I have is of one sunny afternoon after school in late spring, lying face up in one of the fields looking at the clouds, surrounded by cows and cowslips. The sweet, honey perfume and butter-yellow colour of those fairy bells will stay with me for ever; they are my 'scent of memory': Proust has his *petites madeleines*, I have my cowslips.

Wild flower seeds are available in mixes – look out for 'pollen and nectar mixes' or 'bee mixes'. You have to be careful with some of them, however, because they are often formulated for bumblebees rather than honeybees and some are marketed as suitable for separate spring and summer displays. As our aim is to provide food for bees my advice would be to include both spring- and summer-flowering plants in your collection. You can also buy individual species from a number of suppliers: this way you can select varieties to suit honeybees specifically. Some nurseries also stock wild flower plants, which will give you an 'instant' display but can be a little expensive if you have a large area to plant.

Take cuttings

You can increase your shrub numbers by taking softwood cuttings during spring. Another useful way of adding to your stock is to divide some of your summer-flowering perennial plants, particularly those that are getting too big for their allotted space. You can do this at almost any time of year but early spring is one of the best times, as long as the ground is not too wet.

Do some pruning

Early spring is also the time to prune some bee-friendly shrubs, particularly those that flower in late summer. I am thinking of *Buddleja davidii* (butterfly bush) (page 85) and *Olearia* x *haastii* (daisy bush), for example. If left unpruned, *Buddleja* will quickly grow lanky and, since it produces flowers only on new season's growth, it will soon become 'top-heavy' with flowers. Even if we have an early spring and new growth has already started to appear you can still prune the branches hard back, leaving only 5cm or thereabouts of last year's growth. Not only does this keep the bush tidy but also you will be rewarded with a plethora of large flowers – many more than if you had not pruned. The daisy bush is not as unruly as *Buddleja* but you can still prune it in spring to keep its shape. Cut out any dead or diseased wood and prune any other branches back, even into old wood, to promote fresh growth. In fact, it is best to prune young plants back by 50% each year for the first couple of years in order to promote a bushy habit.

Summer – June to August

If spring was the starter then summer is definitely the main course of the annual gardening meal. Much depends on the weather, of course, but given even an average summer, there is a surfeit of food for bees and the nectar cup overfloweth! Now is the time when the majority of plants are flowering – ranging from trees, through to shrubs, herbaceous perennials and fleeting annuals.

June sees the midsummer solstice when the sun is at its maximum elevation and we have the longest day of the year. Bees will take advantage of this and be out at first light, continuing to forage all day until dusk as long as the air temperature is satisfactory and rain keeps at bay. It is no wonder that during the main foraging season worker bees only live for about six weeks: they work every waking hour, flying on average 500 miles in their lifetime.

Buddleja davidii (butterfly bush)

Beware the dog days of late summer though, so called because of the period during which the Dog Star (or Sirius, the 'Shining One') first becomes visible. This star is thought to herald a time of heat and stagnancy, a thunderous, malevolent period, 'when the seas boiled, wine turned sour, dogs grew mad, and all creatures became languid', according to John Brady's *Clavis Calendarium,* 1813. It is now that the nectar flow of many summer flowers starts to diminish and the bees become irritable. Bees do not like close, thundery weather. Despite the gentle smoking, the usual calm thrum of the hive increases in tone and urgency when the bee-keeper opens up the hive for inspection. Bees fly up and tap you on the shoulder as a warning, edging you, persuading you, to move away from the hive. So you take the hint and do not dally, doing what you have to do, and then leaving them in peace.

Summer plants for bees

Trees

Not many gardens are large enough to accommodate a tree, but trees are a mainstay of parks and amenity planting. During the summer months you will find *Castanea sativa* (Spanish or sweet chestnut), *Catalpa bignonioides* (Indian bean tree) (page 87), *Liriodendron tulipifera* (tulip tree) and *Tilia* (lime, or linden) among others, providing varying amounts of bee food. Lime trees were at one time the quintessential street tree but they have fallen out of favour because as well as providing pollen and nectar for bees they can attract large numbers of aphids and scale insects. These suck the sap from the plant tissue, digest it and secrete it as 'honeydew', a sugar-rich sticky substance, which falls on cars parked underneath the trees. Bees sometimes collect honeydew but they prefer to stick with pollen and nectar direct from the flower.

Shrubs

Many shrubs also flower during the summer season. Look out for *Cotoneaster, Lavandula* (lavender) (page 88), *Symphoricarpos* (snowberry), *Perovskia atriplicifolia* (Russian sage) (page 89) and *Buddleja davidii* (butterfly bush) (page 85). You could be forgiven for thinking that *Buddleja* is only suitable for butterflies with their long proboscises, but in fact the nectar

Catalpa bignonioides (Indian been tree)

Lavandula sp. (lavender)

Perovskia atriplicifolia (Russian sage)

rises some way up the narrow corolla tubes allowing honeybees to access it.

Roses

We must not forget *Rosa* (rose) – the nation's favourite flower. Gertrude Stein (the early 20th-century novelist and poet) famously wrote in a non-horticultural context: 'A rose, is a rose, is a rose.' But, horticulturally speaking, roses are not all the same. There are the big, blowsy, old-fashioned blossoms; prim, hybrid teas; multi-bloomed floribundas; repeat-flowering, English roses: these are undoubtedly beautiful (my favourite has to be the English rose 'The Generous Gardener'; not only does it remind me of my late sister (who was not only a gardener *extraordinaire* but generous to a fault in all things), but it has the most exquisite fragrance. Unfortunately, they are of little use to bees because, as we know, double flowers are almost certainly sterile, bearing little or no pollen and are generally devoid of nectar. It is the original, species roses and cultivated single-flowered types which bees need, and because the flowers are not sterile they will produce beautiful hips in the autumn. I have in mind *Rosa forrestiana* (particularly good for bees), *R. macrophylla*, *R. moyesii* and *R. rugosa* (page 91). As well as these, there is *Rosa gallica* var. *officinalis*, the Apothecary's rose, also known as the Rose of Lancaster. Although it is semi-double it does still have pollen. And being a Lancastrian by 'adoption', I just had to include it.

Manuka

In early summer (and, if the weather is clement, as early as May) you will see *Leptospermum scoparium* in bloom. *Leptospermum*, otherwise known as tea tree or manuka, is, for many people, the ultimate bee shrub. (It is not the shrub that tea tree oil comes from, which is *Melaleuca alternifolia*. This is a good example of where knowing the Latin name prevents any confusion.) It is native to New Zealand, where it can be found growing wild, but here in Blighty it is classed as half-hardy, requiring some looking after during the winter months: it can just about cope with a couple of degrees below zero, but having neglected to protect mine last winter I lost it to nine degrees of frost. A lesson learnt. Manuka honey is reputed to be one of the best medicinal honeys and can command a very high premium. (All honey has some medicinal benefits. Both manuka honey and heather honey have been shown to inhibit the bacteria *Staphylococcus aureus* and *Pseudomonas aeruginosa*; manuka honey is reported to have extra benefits. One *teaspoon*

of manuka honey can cost up to £5. You must decide for yourself.) If you keep bees yourself, however, unless you have a plantation of it (as does the Tregothnan estate in Cornwall) then you cannot guarantee that your bees have been feeding solely on that one sort of shrub, so it is best not to get your hopes up about producing your very own manuka honey! Nevertheless it is worth growing one or two of these evergreen bushes simply because bees love them and they are good garden shrubs.

Rosa rugosa (Rugosa rose)

Heather

Heather is the best of all the native shrubs for bees during late summer and early autumn. Commercial bee-keepers will move their bees up to the heather moors where nothing else is in flower during this period because they can guarantee that the heather is the only thing that the bees will feed on and can therefore sell it as pure heather honey. (Bees will fly up to three miles for food but if they can find it on the hive-step then they will not waste energy looking further afield.) There are in fact two shrubs that are known as heather: *Erica cinerea* (bell heather) and *Calluna vulgaris* (ling) (page 93). Bell heather has larger flowers than ling and it is the ling, which flowers slightly later, which produces the 'real' heather honey. This honey is dark with an almost caramel flavour and is often described as the 'Rolls-Royce' of honey. Obviously you will not want to turn your garden into a heather plantation but if you can include some then the bees will love you for it. A word of warning, however; both *Erica* and *Calluna* prefer acid soil, which may pose a problem unless you grow some in a container where you can use specially formulated ericaceous compost for acid-loving plants.

Perennials

Summer is when the perennials really take centre stage. A list of perennials beloved by bees fills several pages, so I shall recommend just a few here (in no particular order of preference) which I can guarantee will keep our buzzy friends happy. *Nepeta* (catmint) (page 7), many *Geranium* (page 76) (not the tender summer 'geraniums', which are really pelargoniums), *Eryngium* (sea holly) (page 94) and *Erigeron* (fleabane) will all get the bees buzzing, as will *Echinops* (globe thistle) (page 94), *Helenium* (page 81) and *Papaver orientale* (Oriental poppies). *Stachys byzantina* (lamb's ears) (page 95) and *Leucanthemum* (perennial daisy) are also worth planting. Bear in mind, too, *Centaurea scabiosa* (knapweed), *Knautia arvensis* (field scabious) and *Achillea millefolium* (yarrow). The last three are wild flowers (although some cultivated varieties of each have been introduced) and are therefore ideal for bees, but they also make beautiful garden plants.

Annuals

Annuals, too, are at their most resplendent now. They take centre stage in summer bedding schemes but if you have a few gaps in your perennial patch it is certainly worth filling them with some annuals. Try *Eschscholzia*

Erica cinerea (bell heather) and *Calluna vulgaris* (ling heather)

Eryngium sp. (sea holly)
Echinops sp. (globe thistle)

Stachys byzantina (lamb's ears)

californica (Californian poppy), *Calendula officinalis* (pot marigold) (page 97), *Helianthus annuus* (sunflower) (page 98) or *Amberboa moschata* (sweet sultan). *Callistephus chinensis* (China aster), *Centaurea cyanus* (cornflower) (page 39), and *Reseda odorata* (mignonette), too, will provide food for bees in varying degrees. Remember though to plant single-flowered varieties.

Patches of annuals can provide a stunning display, too. I visited a garden where a whole border had been given over to mixed annuals (page 99). At first glance it looked like a wild flower meadow, and indeed there were a few native species included in it, but for the most part they were cultivated varieties. The ones which really stood out were variously coloured *Centaurea cyanus* (cornflower) (page 39), *Phacelia tanacetifolia* (page 101) and *Papaver rhoeas* (cultivated corn poppies, probably 'Mother of Pearl') (page 101), all of which were proving to be magnets for both honeybees and various bumblebees. The seeds had obviously been sown directly into the soil and been left to their own devices; the result was spectacular.

Beware the thug!

One perennial plant which is extremely attractive to bees, but is an environmental calamity, is *Impatiens glandulifera* (Himalayan balsam). It was introduced to Britain in 1839 but it has escaped from the confines of the garden and has naturalized, especially along riverbanks, where it has become a major problem. Aesthetically it is a lovely plant, but environmentally it is a thug. It grows rapidly and spreads quickly by means of 'explosive' seed pods producing up to 800 seeds each, which can be shot up to seven metres.

I am not saying you should not grow it (especially if you want to attract bees) but you must be very careful to keep it in bounds and be vigilant not to let it spread seed. If you want to grow it year on year, then allow just one or two flowers to produce seed pods and before they ripen and have a chance to explode, enclose them in strong paper bags (not plastic ones). This will allow the seed to ripen, but will prevent it from spreading if you do not manage to harvest it before it explodes. Keep an eye on the weather, too, and replace the bags if they get soggy in the rain.

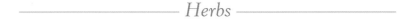

Herbs

Herbs are mainly grown for their leaves but if you can let one or more plants flower then you will provide a welcome variation to the bees' menu. (And

Calendula officinalis (pot marigold)

to your own, since a number of herb flowers, such as *Borago officinalis* (borage) (page 24), *Allium schoenoprasum* (chive), *Eruca vesicaria* var. *sativa* (rocket), *Myrrhis odorata* (sweet cicely) and *Lavandula* (lavender) (page 88) can be eaten and are a delicious addition to a summer salad.) I mentioned earlier that one of the best bee herbs is *Origanum vulgare* (marjoram) (page 25), the flowers of which produce the most concentrated nectar known, containing 79% sugar. *Allium schoenoprasum,* (chive), *Hyssopus officinalis* (hyssop) (page 67), *Mentha* (mint) (page 68), *Salvia officinalis* (sage), *Satureja hortensis* and *S. montana* (summer and winter savory) and *Melissa officinalis* (lemon balm) all flower at this time of year.

Lavender

I think, though, if I were forced to choose just one herb to grow it would be *Lavandula* (lavender) (page 88). (The species *Lavandula angustifolia* and the hybrid *L.* x *intermedia* are the hardiest.) First of all it is a superb bee plant, which is good enough reason (and lavender honey is sublime), and it is easy to grow. Lavender flowers can be used in cooking, in both savoury

Helianthus annuus (sunflower)

A border of mixed annuals

and sweet dishes. I combine a few lavender flowers with chopped thyme and tarragon, then add some olive oil to use as a rub on a leg of lamb before roasting it; and I make scones with just a few lavender flowers in the mix – cut open and spread with butter and honey the scones are just scrummy. Still in the home, you can use dried flowers in lavender bags to fragrance linen and to deter moths.

Lavender also has medicinal uses, but you must be careful to get expert advice before you use it: the essential oil has antiseptic and anti-inflammatory properties and is reputed to aid sleep and relaxation. And, finally, it is found in many perfumes. There are very few plants that can claim to do all those things and still look beautiful! My favourites are *Lavandula angustifolia* 'Hidcote', a compact shrub with dark purple flowers, *L. angustifolia* 'Maillette' which is a little larger, has mid-purple flowers and is the most widely grown for oil, and *L.* x *intermedia* 'Sussex' which is larger still and produces pale purple flowers over the longest period of any of the hardy lavenders. (I have to admit a slight bias in my last choice because I was born in Sussex.)

Other plants

Some summer-flowering plants which are grown as agricultural or horticultural crops and are sought after by commercial bee-keepers because of their high nectar or pollen yield can also be useful in the domestic garden. I am thinking particularly of *Borago officinalis* (borage, also known as starflower) (page 24), *Helianthus* (sunflower) (page 98), *Onobrychis viciifolia* (sainfoin), *Trifolium* (clover – yes, I know it is thought of as a weed) and a whole range of fruit and vegetables.

Borage is classed as a herb and is grown commercially mainly for its oil but the flowers are lovely in a salad and you cannot have a proper Pimm's in the summer without its leaves. Sunflowers are grown commercially for seed and oil, and sainfoin is making a comeback as a fodder crop. Farmers will grow clover to increase the nitrogen levels in the soil but for the average gardener clover is anathema. Forget your prejudices because *Trifolium repens* (white clover) (page 25) is one of the best bee plants known. Another plant, *Melilotus officinalis* (page 103), whose common name is sweet clover or melilot, is of the same family as, but a different species from, white clover. It looks quite different from 'ordinary' clover in that it grows to a height of 30–50cm and has pea-like yellow or white flowers. It rates as highly as white clover in the bee plant chart.

Phacelia tanacetifolia

Papaver rhoeas (poppy)

Edible plants

I have already mentioned that fruit and vegetable yields can be increased by anything up to 36% if bees are active on the crop. This is a huge advantage for farmers and growers. And if it is true for commercial crops then it is also true for domestic ones. Summer fruits, such as strawberries, gooseberries, raspberries, blackberries, currants – indeed all soft fruits – will have an increased yield if pollination is high. Fruiting vegetable crops, such as broad and runner beans (page 104) will benefit (but not peas which are self-fertile and do not need external pollination), as will the cucurbit family of courgettes (page 105), marrows and pumpkins.

Summer jobs for bee-friendly gardeners

Summer is one of the busiest times for the gardener but if you don't get around to doing every single job then it doesn't matter – there is no point getting upset. Gardening should be a pleasure, not a chore, and unless you totally neglect your plot for months on end then the worst that will happen is that the garden will begin to look a little unkempt. Never mind.

If you do have time, however, there are jobs to be getting on with. There's weeding, of course, but let's not say any more about that!

Planting and sowing

The risk of frost has long passed and the soil has warmed up enough to start planting out the more tender flowering annuals that you so assiduously sowed under cover earlier in the year and which you have been potting on since. Early summer is also the time to sow some annuals directly into the ground. Many annual seeds prefer this method but, once sown, they dislike being disturbed. You must thin them out, of course, otherwise the growing space will become cramped and the plants will grow leggy and weak, but apart from this, leave them alone and they will give you a beautiful display with a minimum of effort on your part.

Many vegetable seeds can be sown now, too, either directly into the ground or, if there is not enough room because crops haven't quite matured, they can be sown in pots or modules, ready to be planted out later.

Visit your local nursery – again

If you have bigger gaps in your planting than a few annuals can fill, then

Melilotus officinalis (sweet clover)

Runner bean

pop along to your local nursery to stock up. The beauty of containerized plants is that you can plant them at any time of the year, except in the dead of winter if the ground is frozen or is too wet, so you can see them in flower at your local nursery before you buy. If you do plant anything during the summer, however, make sure that the ground is well prepared and that you water the plants diligently during the first season. Thereafter they should be able to fend for themselves.

Take some cuttings

During late summer and into autumn you can increase your stock of all sorts of plants, ranging from trees and shrubs to climbers and herbs, by taking semi-ripe cuttings.

Deadheading

One more job that needs to be done throughout the summer and into autumn is deadheading the flowers. This is simply removing the spent flower heads once they have bloomed. The main reason for doing this is to

Courgette

encourage the plant to direct its energy into producing even more flowers. If you do not deadhead then the plant's energy will be used in forming seed heads, capsules or pods, which will then be shed and, assuming they are viable, will give you scores or even hundreds of seedlings germinating around the parent plant. If you don't mind this too much you can, towards the end of the season, leave some seed heads on the plants to provide food for birds over the colder months. The main candidates for deadheading are annual and perennial flowers, but don't forget shrubs, too. Some shrubs will often give a second flush of flowers if you remove the first after flowering. But if you want hips on your roses, do not deadhead them!

Do some more pruning

Alongside deadheading there is also some pruning that can be done in summer. I am thinking particularly of some fruit trees, especially espalier-, cordon- or fan-trained *Malus* (apple) and *Pyrus* (pear) and *Prunus* (plums and other species). Have a look on the website of the Royal Horticultural Society for advice on how to prune these trees. Some spring- and summer-flowering plants can be pruned or cut back after they have finished flowering. *Lavandula* (lavender) (page 88), that well-beloved bee plant, is a prime candidate for this treatment. It should be pruned every year to keep it nice and compact. When you do prune, remove the flower stalks and about 2.5cm of the current year's growth; always leave some current growth because lavender does not regenerate easily from old wood.

Do not forget ...

And finally, there is something at this time of year which too many gardeners forget to do. Imagine you have weeded and watered, mown and mulched, deadheaded the flowers and done similar things to slugs. What is there left to do? Nothing. Nothing? Just so. Take some time out. Sit in the garden. Look at the flowers. Listen to the bees. Just be. And enjoy.

Autumn – September to November

My spirit sinks a little in September when I see the house martins and swallows line up on the telegraph wires: I am being selfish, willing them to stay just a little longer, but instinct tells them they must be gone. September also sees the autumn equinox when day and night are of equal length, and the harvest moon (the full moon nearest to the autumn equinox) is often, but not exclusively, seen in September. (At the time of writing, 2010, these events actually occurred on the same date.) The harvest moon is so called because farmers would take advantage of the brightness of the moon to work well into the night, bringing the harvest home.

Autumn is truly the time when nature provides an abundance of good things and I cannot help but think of Keats's well-loved poem, 'To Autumn', that 'season of mists and mellow fruitfulness', providing:

> … *more,*
> *And still more, later flowers for the bees,*
> *Until they think warm days will never cease;*
> *For Summer has o'erbrimm'd their clammy cells.*

The bees have already feasted on the summer bounty and now autumn presents the dessert: not as opulent as the main course of summer perhaps but still worthwhile. As the days get shorter and the temperature falls the bees will take every opportunity to forage for pollen and nectar in order to fill the combs to the brim with honey reserves for the winter. (And then the bee-keeper comes along and harvests the honey – and very nice it is on buttered toast in the morning.)

Although the variety of flowering plants is not as great as earlier in the season, there are still some worthwhile bee-friendly ones that we can look out for during the autumn – and things that we can do in the garden.

Autumn plants for bees

Trees

Autumn-flowering trees are, it has to be said, few and far between. There are no indigenous trees in flower at this time of year but there are some non-native ones that are less widely grown and are very good 'bee' trees. These include *Sophora japonica* (pagoda tree) and *Tetradium daniellii* (the appropriately named Chinese bee tree). The latter, especially, provides a profuse nectar supply when other sources are waning.

Shrubs

There are some shrubs that are worth looking out for too. *Fuchsia* (*F. magellanica*) (below), some varieties of *Hebe*, *Perovskia atriplicifolia* (Russian sage) (page 89) and *Viburnum tinus* (page 73) will provide some pollen and nectar. *Calluna vulgaris* (ling heather) will also flower into the autumn. Another shrub which is well worth growing if you have the space is *Eucryphia* x *nymansensis* 'Nymansay' (leatherwood) (page 109). I came first came across this shrub in full flower in a friend's garden in early September. It was the sight of beautiful white flowers which first attracted me, but as I grew closer it was the sound which dominated. There was a constant, gentle hum of honeybees diligently working the flowers. I have never seen so many bees

Fuchsia magellanica

on one shrub – admittedly it was a very large shrub, having thrived in its sheltered position within a walled garden – but the number of bees in relation to the shrub's size was still astounding.

Ivy

One of the most useful bee shrubs to have around in mid-to late autumn is undoubtedly the climber *Hedera helix* (ivy) (page 110). This flowers at the very end of the season and is an extremely important source of nectar, containing about 70% glucose. This is of particular value if we have good weather in October and November and the bees are still foraging. Nonetheless, most people are reluctant to plant it in their garden because

Eucryphia x nymansensis 'Nymansay' (leatherwood)

of its assumed invasive nature and the damage it allegedly does to masonry.

It is true that, with patience, it can take on enormous proportions but there is little conclusive evidence that it is the ivy that causes brickwork and the like to fail. However, given that the flowering heads which provide the nectar and pollen in which bees are interested only appear on 'adult' growth, which can take 15 years to reach, it is not really worth growing it in a new garden. If it is already there, then leave it; otherwise let the bees venture into the parks and countryside to source it.

Perennials

Many summer-flowering perennials, if they are deadheaded, will put on another, albeit less flamboyant, display during the autumn until the first frosts.

Hedera helix (ivy)

Dahlia sp.

In addition there are those which save their best until now, like *Aster* (Michaelmas daisies) (page 82), *Dahlia* (but make sure they are the single-flowered varieties) (page 111), *Rudbeckia* (coneflower) (page 31) and *Centranthus ruber* (valerian). Another useful source of autumn food, and one which may not spring readily to mind, is *Colchicum autumnale* (autumn-flowering crocus or meadow saffron) (below). The leaves appear way back in spring and have died down long before the naked flowers reveal themselves.

Colchicum autumnale (autumn crocus)

Autumn jobs for bee-friendly gardeners

The days are getting shorter, the bees are becoming less active, and the garden seems to be starting to wind down after the hectic ostentation of the summer months. There are still jobs that the gardener can be doing in autumn, though.

Lift tubers

When the first frosts come and blacken the foliage of the dahlias then it is time to lift the tubers (if you live in a fairly mild part of the UK then you could risk leaving them in the ground over winter, covered with a good mulch of straw, but living in the north-west I have never dared to). Cut off all the stems, leaving just a few centimetres, then dig up the tubers, taking care not to damage them, and shake off as much soil as you can. Rinse off the rest of the soil and stand the tubers upside down to dry. When they are completely dry put them in a box or large pot and cover them with *dry* multi-purpose compost, shaking the box to make sure that the compost comes into contact with the tubers. Then put the box in a frost-free place – this is vitally important because if the tubers freeze they will not recover. They will happily lie dormant until next spring when it is time to bring them out of 'hibernation'.

Mow the meadow

If you have a wild flower patch with plants that predominantly flower in the summer, early autumn is the time to mow it, once wild flowers such as *Succisa pratensis* (devil's bit scabious) and *Centaurea scabiosa* (knapweed) have set seed. The vegetation should be cut to a height of about 7.5cm and left for three or four days to allow any remaining seed to be shed. The mowings must then be cleared away: if they are left *in situ* then the dying vegetation will smother existing plants. In addition, the debris will rot down and add nutrients to the soil, increasing its fertility, which is just what you want to avoid since wild flowers fare best on poor soil.

Plant spring-flowering bulbs

One of the main tasks in autumn is to plant bulbs which will provide some food for bees next spring and early summer. *Anemone blanda* (windflower), *Muscari* (grape hyacinths) and *Tulipa* (pages 74–5) all provide pollen for bees just at the time when the brood is starting to build, but little in the way of nectar. *Eranthis hyemalis* (winter aconite) and *Crocus* provide both pollen and nectar as do ornamental alliums such as *Allium hollandicum* 'Purple Sensation' (page 115), a stalwart of Chelsea Flower Show gardens, which look sensational (excuse the pun) in the herbaceous border.

Most bulbs need to be in the ground by the end of September, although tulips are best left until October or even November because they are susceptible to a fungal disease called tulip fire, which thrives in warmer weather. All bulbs should be planted three times their own depth so if you have a 5cm bulb then it should be covered with 10cm of soil.

A friend of mine who has a large garden (lucky person!) planted several thousand bulbs a few years ago. The first spring they flowered well, but thereafter the flowering rate diminished until this year all that appeared were leaves and a few scrawny flowers. She was bitterly disappointed. She had bought the bulbs from a reputable source and had done everything according to the book – except one thing. She hadn't planted them deeply enough. Instead of planting them three times their own depth she had just pushed them into her lovely loam soil and scraped a bit of earth over the top. It was a costly mistake in terms of time, effort and money. If you think you are planting too deeply, you are probably planting just deep enough. You can also plant up some containers with bulbs, either to have as a feature in their own right, or to slot into your border to provide a bit of colour for next spring.

Plant spring bedding

As well as bulbs you can also plant out some spring and early summer bedding, such as sweetly scented *Cheiranthus cheiri* (wallflowers) (page 77). Although you can raise these from seed sown in the spring, they are widely available at nurseries and garden centres. Do not bother buying them in pots. They are much cheaper bare-rooted and will come to no harm as long as you soak the roots well in a bucket of water before you plant them out.

Collect and sow seeds

In early autumn you can collect seeds from perennials, hardy annuals and half-hardy annuals and sow them under cover. This will give them a longer growing period and a head start over similar plants sown in the spring, so they will flower earlier.

Allium hollandicum 'Purple Sensation'

Take some cuttings

If you have some tender perennials in the garden, like *Verbena rigida* (page 59) or *Penstemon* (page 36), autumn is a good time to take cuttings to make sure you have some plants for the following year. If there's a mild winter, then the current year's plants may well survive, but several good frosts are likely to see them off. Follow the procedure for taking softwood cuttings, but make sure you keep them indoors, out of the cold, at all stages until the following year when they can be planted out in the garden when all sign of frost has gone.

From mid-autumn until well into winter you can take hardwood cuttings to increase your stock of bee-friendly deciduous shrubs, such as *Buddleja davidii* (butterfly bush) (page 85) and *Ribes* (flowering currant) and some bush fruits, such as gooseberries, and black, red and white currants. The second half of September is also a good time to take hardwood cuttings from roses. There are a few evergreen shrubs from which you can take cuttings at this time of year as well, notably *Cotoneaster*, *Ilex aquifolium* (holly) and *Skimmia*. You should treat these like semi-ripe cuttings.

Another method of getting some plants free is to take root cuttings. This can be done from mid-autumn through to early winter and suits some herbaceous plants which have thick or fleshy roots such as *Papaver orientale* (Oriental poppy), *Echinops* (globe thistle) (page 94) and *Echinacea purpurea* (coneflower) (page 82).

Mulch

Mulching can also be done now. The idea of mulching in the autumn is to trap the warmth and moisture in the soil. Therefore it is best done before the first frost arrives but do make sure the ground is not dry – any rain falling after mulching will have another three or four inches to penetrate before it gets to the soil proper and will do little to sustain the plants.

Winter – December to February

By December the first frosts will have arrived and we may even have had some snow. I don't really mind the snow simply because we don't usually get very much of it and when we do it doesn't lie around for too long. Some winters, however, turn out to be the exception that proves the rule, bringing snow and well-below freezing temperatures for more days than we are used to. When this happens, I can't help but think of the painting by Monet entitled 'The Magpie', which shows a wintry landscape, a hedgerow topped with a thick ridge of snow and a solitary magpie perched on a gate – a painting which can be found on many a Christmas card. I do wonder if there are any beehives tucked behind the hedge, sheltered from the worst of the wind and snow.

The garden doesn't seem to mind the snow that much, though. It acts almost as a blanket, protecting the soil and plants underneath it from dramatic changes in temperature. Honeybees are not bothered too much about the snow either. They will be intent on keeping the queen alive. Even so, you may see the odd bee flying on a warmer, bright day just to see what's going on and whether there is anything worth reporting back about. The critical temperature for individual honeybees is 6°C. If they do venture out of the hive and the temperature falls below this level, they will die. (Bumblebees, however, can and will fly at much lower temperatures.) There are some plants which flower during the winter months but hopefully our bees will recognize that they are just tasters of things to come, and are not part of a proper menu.

Winter plants for bees

Winter-flowering plants are few. Winter-flowering plants that are suitable for bees are even fewer: the sweet-smelling *Viburnum x bodnantense* is one such and *Mahonia aquifolium* (Oregon grape) can be relied upon for a valuable pollen source during this season. *Erica carnea* (winter heather) will also flower now as will *Ulex* (gorse). It is not really surprising that so few plants are flowering at this time of year – most insects are hibernating or in a state of near torpor, which nature is well aware of, so there is not the same imperative to provide food.

Winter jobs for bee-friendly gardeners

Gardeners, too, are often in a state of torpor, especially if they have imbibed too much over the festive period. It does not matter so much, however, because the number of jobs that can be done during the winter period is relatively small.

Plant trees and shrubs

As long as the ground is not frozen and is not too wet, bare-root trees and shrubs, especially roses, can be planted during the dormant period, which is usually from November to March. You can also plant containerized plants at this time but, because they are in pots, you can actually plant them at any time during the year as long as the conditions are favourable.

Take cuttings

In the field of horticulture nothing is more rewarding than getting something for nothing so continue to take hardwood cuttings.

Do some pruning

Roses can be pruned during the winter. Hopefully, you are growing bee-friendly ones in which case you may want to make the most of the beautiful hips and delay the pruning until late winter. Pruning these roses is very simple; all you need to do is to remove dead or damaged wood and, if you want to restrict the size, cut back by no more than a third and thin out the branches very lightly. That's it.

Take stock

Now is also a good time to take stock and see if any alterations can be made to an existing garden to encourage bees, or, if you are in the enviable position of planning a new garden, to decide what should be included to bring bees and other beneficial insects into your space. Have a look at the suggested planting plans later in the book, look through this chapter, or pick out some plants from each of the seasons in the gazetteer to include in your garden, and then consult the numerous seed catalogues that land on the doormat soon after Christmas.

My plantaholic desires have taken many a flight of fancy while I have been poring over these seed and plant lists: I am seduced by rich, dark 'Queen of the Night' tulips and 'Midnight Blue' agapanthus; enticed by confectionery colours of candy-pink morning glory, chocolate cosmos and lemon sherbet dahlias; covetous of jewel hues of ruby-red astrantia, sapphire-blue cornflower and opalescent iris.

But I am soon brought forcefully down to earth again when I tot up how many seeds, bulbs and plants I have earmarked and, more to the point, how much they will cost. My long-suffering husband invariably points out, too, that our garden would have to be the size of a minor principality to accommodate all the plants that I have in mind. Ah, well, that's the beauty of armchair gardening – the only limitation is the capacity of your imagination. So I shall doze off, seed catalogue in hand, tea at my side, and dream of balmy summer days when the sun is shining, the flowers are blooming, and my bees are buzzing happily around me …

4
Other Flying Insects

The honeybee is often mistaken for a bumblebee, solitary bee, hoverfly or wasp, so I would now like to look briefly at these other important pollinating insects, all of which will be attracted into your garden even if you have primarily concentrated on providing food for honeybees.

Bumblebees

Colony-forming bumblebees

Let's start with colony-forming bumblebees. Although we call them colony-forming, these bumblebees do not live in 'families' numbering thousands, nor do they build comb, like the honeybee. The queen bumblebee has quite a different role from that of the queen honeybee, too.

In all colony-forming bumblebees, only queens (fertilized females) survive over the winter months; all the males will have died the previous autumn. The first thing the emerging queen will do is to find food and then a suitable nest site in which to lay the fertilized eggs that she has carried inside her throughout the winter hibernation. Where she chooses to build her nest depends on what species of bumblebee she is and we shall look at individual species shortly. Once she has built her nest she will collect together a ball of pollen mixed with a little nectar to 'glue' it together and will coat it with wax. She will also make a small wax cup, set it next to the pollen ball and fill it with nectar. With these preparations completed she will then lay batches of eggs and press them gently into the pollen ball.

Unlike the queen honeybee who, once she has laid her egg, leaves all the nurturing to worker bees, the queen bumblebee will 'sit' on her eggs and incubate them much as a bird does. This is where the cup of nectar comes in handy to sustain her through the incubation period. Once the eggs hatch they start to feed on the pollen ball but she then has to venture out of the nest to collect even more food for her babies. After about 14 days the larvae pupate and after a further 14 days adult bees emerge from the cocoons, much as butterflies do. The first brood of bees will invariably be sterile female worker bees. It is now their job to take over the foraging and feeding duties so that the queen can concentrate on laying further batches of eggs.

In early summer the queen will start to lay eggs that will emerge as either male, or female bees which will develop into future queens. No more female worker bees will be produced. By late summer the old queen and any remaining worker bees will die. The new queens and males emerge from the nest and mating takes place with bees from different colonies. After mating, the new queens will find a suitable place in which to hibernate and the males will die off.

Most common bumblebees

There are six main species of colony-forming bumblebee that we are likely to come across in our garden.

Buff-tailed bumblebee

Usually the first bee to emerge in spring – often as early as February – is *Bombus terrestris*, the buff-tailed bumblebee (opposite), and you can guarantee that this early in the year it will be a queen. This bee has a yellow/buff stripe on the thorax, a similar-coloured stripe on the abdomen and a light buff tail and is between 10 and 16mm long. She will often choose a hole in a bank or an old mouse hole in which to build her nest and before long the colony will have built up into about 150 bees. Although they are quite large, buff-tailed bumblebees are fairly harmless, and will only sting you if they feel under threat.

White-tailed bumblebee

You may also see *Bombus lucorum*, the white-tailed bumblebee, which is very similar to *Bombus terrestris*, with quite similar markings apart from her tail, which is white rather than buff. She is also about the same size, namely 10–16mm long.

Early bumblebee

Next is *Bombus pratorum*, the early bumblebee. This bee is smaller than *Bombus terrestris*, being 9–14mm long, and she has slightly different markings: a yellowish stripe on her thorax, a buff-coloured stripe on her abdomen and an orange/red tail. She too builds a nest, though often in unusual places like nesting boxes. I wonder if, having set up home in a bird box, this little bumblebee has adopted the habit of some birds to have more

Bombus terrestris (buff-tailed bumblebee)

than one lot of offspring in a season, because it has been reported that queens from a new brood will often start another colony straight away instead of hibernating. As far as foraging goes, *Bombus pratorum* feeds on the same sort of flowers as honeybees, so you may well find them sharing a bloom.

Red-tailed bumblebee

Bombus lapidarius, the red-tailed bumblebee (page 124), is probably the most easily recognized species with its black body and bright orange tail. This bee is about 12–16mm long. You will find these bees nesting underground and in the base of dry-stone walls. The size of the nest can vary considerably from over 200 bees to fewer than 100.

Bombus lapidarius (red-tailed bumblebee)

Common carder bee

Bombus pascuorum, the common carder bee (page 125), is the most widespread of the bumblebees and can be found just about anywhere in the country. It probably has the least spectacular colouring of all the common bumblebees, having an orangey/buff thorax and buff abdomen, and is probably overlooked by many. This bee can be between 10 and 15mm long. Some people say this bee can be a little feisty if you disturb it at home in its tussock of grass, but I don't think I would be too pleased if the equivalent of a size 130 boot landed on my bedroom ceiling either!

Garden bumblebee

The last of the bumblebees that you are most likely to see in the garden is,

Bombus pascuorum (common cader bee)

funnily enough, the garden bumblebee, or *Bombus hortorum*. This bee has very similar colouring to *B. lucorum*, in that she has a yellowish stripe on her thorax and a white tail. The main difference is that she has a double yellowish stripe on her abdomen. Like *B. pratorum* she is often found nesting in odd places such as inside lawnmowers or the pocket of your old gardening coat that you leave in the shed. The colonies are quite small and short-lived, so do not disturb them unless you really have to.

Solitary bees

In addition to bumblebees you may encounter one or more of the 250 species of solitary bees to be found in Britain. Solitary bees are just what their name implies: they live and work alone rather than within groups as do bumblebees, which share nests and work together to raise offspring. Some of the most common solitary bees are mining, mason and leafcutter bees.

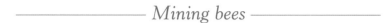

Mining bees

Female mining bees (*Andrena* sp.) dig long, narrow burrows with branching cells in the ground – often in lawns. (You may see a small pile of excavated soil at the entrance to the burrow.) At the bottom of each cell the female will deposit a clump of pollen and nectar that she has taken up to seven journeys to collect, and then lay one egg. When the larva hatches it will eat the store of food and then pupate over winter. The following spring the bees emerge, mate, and the cycle begins again. They are very docile bees and seldom, if ever, sting.

Mason bees

One of the most common mason (or masonry, or mortar) bees is *Osmia rufa*, or the red mason bee. This bee is opportunist and will build nests in existing hollows such as gaps in plant stems or mortar joints. Interestingly, the bee does not get its name from choosing brickwork or stone for its nest; it comes from the fact that it will separate cells within the nest hollow with a wall of mud. Like the mining bee, the female will lay one egg in each cell, provisioned with pollen and nectar for the larva which will emerge the following spring as an adult bee. This bee will sting you, but apparently only if you squeeze her between your fingers!

Leafcutter bees

Another solitary bee that you might come across is *Megachile centucularis*, the leafcutter bee. You will know that you have this bee in your garden by the telltale semicircular chunks you will find munched out of the edges of leaves, particularly those of roses. The female bee uses these segments to construct and seal the cells in each of which she lays one egg. A nest will consist of six to ten cells, and as each cell takes about eight hours to construct, it really is a labour of love. Leafcutter bees will nest in soft, rotted wood or similar material that is easy to excavate, so if you really do not want them in your garden because of what they do to your roses, just make sure that there are no suitable nest sites and that should do the trick.

Hoverflies

Another flying insect that will undoubtedly visit your garden is the hoverfly, which belongs to the family *Syrphida* (page 128). As with the bumblebee, there are numerous different species to be found in the United Kingdom (more than 230 in fact) so the chances of avoiding one are pretty remote! They do not sting, so to try to avoid becoming something else's dinner, they are masters of disguise: many have the colouration of wasps or bees and some are even hairy, like bees. Unlike bees, however, hoverflies have only one pair of wings – and, not surprisingly, they have the ability to hover. Another difference is that whereas bee larvae are fed pollen and nectar, hoverfly larvae are fed on aphids, so they are really good guys to have in the garden. Depending on the species, hoverflies 'nest' in a wide range of habitats: some will use decaying wood, others have adapted to aquatic life; what they do not do is live in colonies like honeybees.

Wasps

When it comes to flying insects it is their ability to sting which is really the object of our fear. Unlike bees, whose stings are barbed and can only be used once, wasps have no barbs on their stings and can therefore use them repeatedly. And this is what they will do without any compunction – or so it seems to us humans! I was going to say, the least said about this little creature, the better. But people so often mistake the honeybee for a wasp and start splatting it, that it is worth spending a few minutes looking at my least favourite insect.

There are seven species of wasp to be found in the UK but the two you are most likely to come across are *Vespula vulgaris*, the common wasp (page 129) and *Vespula germanica*, the German wasp. They both have the typical wasp features of bright yellow and black stripes on the abdomen and a pinched-in waist (hence the term 'wasp waist'). The only obvious differences between them are that *Vespula germanica* has three black dots on its face and the yellow stripes on its abdomen are more pronounced.

Like the bumblebee they build a new nest each year after the queen has emerged from winter hibernation. They do have a role to play in the garden in that wasp larvae (like hoverfly larvae) eat animal (or rather insect)

Hoverfly, a member of the Syrphida family

protein in the form of aphids and suchlike, which we would often consider to be pests. Adults feed on nectar and other sweet substances, which is why we will find them making a nuisance of themselves, buzzing around our fizzy drinks, jam tarts – and honey.

Their predilection for honey and insect protein is what makes a bee-keeper like me so 'anti-wasp'. Wasps will happily 'invade' hives, killing the sentinel bees guarding the entrance to the hive, to steal the honey stored inside. In addition, and what is worse, they will also take bee brood (bee larvae and pupae) from inside the hive and will attack and abduct foraging worker bees to feed to their own larvae. Not nice.

Vespula vulgaris (common wasp)

Plants for bumblebees

I have said elsewhere that if you provide suitable food for honeybees then you will also be helping other insects. This is undoubtedly true when it comes to the ones we have been looking at here. All of these insects will appreciate every single honeybee-friendly plant you have in your garden. If you have a particular penchant for bumblebees, however, and particularly those with a longer proboscis like *Bombus hortorum* (the garden bumblebee, whose proboscis is some 13.5mm long), you may like to grow some plants whose flowers are more adapted to pollination by bumblebees rather than honeybees.

I am thinking particularly of plants with long corolla tubes, like the clovers, *Trifolium pratense* (red clover) and *Trifolium hybridum* (alsike clover) (page 131), as well as *Buddleja* (butterfly bush) (page 85) and *Jasminum* (jasmine). *Monarda* (bee balm), too, has long corolla tubes which even some bumblebees would struggle with. Despite its common name the nectar of cultivated varieties (which are hybrids of *Monarda fistulosa* and *M. didyma*, whose respective average corolla tube lengths are 7.5mm and 15mm) is only accessible to a few bees.

There are flowers which have 'lips' too, which only the bumblebee has sufficient weight to prise open to get at the nectar. These include *Aconitum* (monkshood) and *Antirrhinum* (snapdragon) (page 41). In addition there are those plants which have 'trumpet'-shaped flowers which, although not inaccessible to honeybees, are easier for bumblebees to access, like *Cerinthe major* var. *purpurascens* (honeywort) (page 132) and the bumblebee plant *par excellence*, *Digitalis* (foxglove) (page 133). I have some of the latter in my garden and whenever I look there is at least one bumblebee making an expedition into the far recesses of the flowers, and then backing out in a rather ungainly fashion.

Homes for bees

Although you may not necessarily want to have a honeybee hive in your garden, it is fairly easy to provide shelter and accommodation for the other insects I've mentioned, especially solitary bees and bumblebees. Here are a couple of suggestions.

Trifolium pratense (red clover) *and Trifolium hybridum* (alsike clover)

Solitary bee house

You can buy specially-made solitary bee 'houses' at garden centres or online, some of which are very attractive. You can also make your own quite easily and, although it might be less attractive, it will fulfil its purpose. The easiest method is to take a block of untreated timber, about 12–15cm long, and drill a series of holes in the end of it with diameters of between 2mm and 10mm. Don't drill right through, though – you want a channel which is a cul-de-sac, not a through road. Clear away any sawdust and make sure the edges of the holes are smooth and that there are no splinters.

Another example, which is a little more attractive, is to take a log, again about 12–15cm long, and drill or carve out the inside, leaving a solid back

Cerinthe major var. *purpurascens* (honeywort)

to it. Then fill the space with lengths of bamboo cane of differing diameters, again making sure that there are no rough edges where you have cut them. Also make sure that they are well packed into the log and do not rattle around at all (page 135).

Whatever form your bee house takes, find somewhere sheltered to put it which is at least a metre off the ground. It should also be in full sun, facing south or south-east – don't forget bees are cold-blooded and rely on the sun to warm them up. If you want to do so, you can make a bee 'apartment' block by stacking several houses on top of one another.

You may find that other insects such as lacewings or ladybirds take up residence in some of the holes in your bee house, which is a bonus.

Digitalis purpurea 'Alba' (foxglove)

Bumblebee house

If you want to provide a home for a bumblebee then, again, you can buy a commercially-produced one, usually made from wood. You can also make your own but, rest assured, you do not need any carpentry skills. All you need is: a flowerpot (preferably frostproof terracotta) at least 20cm in diameter; a piece of slate or tile big enough to cover the hole in the bottom of the flowerpot; a length of flexible pipe, about 30–50cm long and at least 18mm in diameter (you will need to put drainage holes in the pipe – how you do this will depend on the pipe material); a piece of chicken wire, big enough to make a 'cradle' to put inside the pot; a handful or two of dry moss, or pet bedding – hamster bedding is ideal. Now all you need to do is put it all together.

Dig a shallow hole in the ground and place the hose so that one end is at ground level and the other is within the area that you have excavated. Place the 'cradle' of chicken wire in the centre and fill it with moss or bedding – this is the nest. Then place the upturned flowerpot over the nest, back-filling around the bottom so that the only way into the nest is via the hose. Pop the slate or tile on top of the flowerpot to cover the drainage hole so that no rain can get in. And there you have it – a bijou detached des res for a bumblebee!

A home for solitary bees (opposite)

5

Gazetteer of Plants Attractive to Bees

Like humans, bees like a varied diet and therefore will not restrict themselves to one kind of plant. In this chapter we will be looking first at general plant families that are particularly attractive to bees. This is really just a 'drop in the nectar ocean' as there are over 500 different families ranging from *Acanthaceae* to *Zygophyllaceae*, many of which contain bee-friendly plants, even if it is only one or two. To look at each in turn is beyond the scope of this book, so I have concentrated on families which contain a significant number of bee plants.

After that you will find sections, each dealing with a type of plant (annuals, perennials, shrubs, trees, edible plants, herbs and wild flowers), arranged more or less in order of the season in which they start flowering. (I say 'more or less' because flowering times will vary depending on the weather, location and so on, so they are approximations rather than certainties.)

In almost every section you will also find flowering times in the form of a chart, which I hope will be useful as a quick reference guide. The lists of plants you find in these sections are certainly not exhaustive, but they will give an idea of the sorts of flowers that keep cropping up when we think of 'bee-friendly' ones.

Plant families

Lamiaceae

First we have *Lamiaceae* (formerly known as *Labiatae*). This is the 'family' name and contains a whole range of plants running into hundreds of species. Those which are of most interest to us are common ones such as *Lamium* (dead nettle) (page 211), *Lavandula* (lavender) (page 88), *Rosmarinus* (rosemary), *Mentha* (mint) (page 68), *Origanum vulgare* (marjoram) (page 25) and *Salvia* (sage). They are often aromatic (many are culinary herbs) and are easy to cultivate. The flowers typically have petals fused into an upper and lower lip (hence the original family name) and are positioned in clusters around the stem. This means that the bee can visit scores of individual flowers on one stem to collect nectar, using very little energy and pollinating numerous flowers at the same time.

Asteraceae

Like the *Lamiaceae* family, *Asteraceae* (once known as *Compositae*) contains a huge number of species (nearly 23,000 currently accepted species). Many flowers look like daisies but the family name comes from *aster*, which is Greek for star. The main characteristic of this family is that the flower 'heads' are made up of many individual flowers. These individual flowers can be regular, with all the petals the same size, often forming corolla tubes (like *Taraxacum officinale* [dandelion]). They can also be irregular, with some petals bigger than others; the bigger petals usually surround the central disc of short flowers: a good example is *Helianthus annuus* (sunflower) (page 98). Many members of the *Asteraceae* family produce quantities of nectar and pollen which are easy to access – and there is a 'landing platform' to boot! – so they are ideal 'bee plants'.

Rosaceae

The *Rosaceae* family contains not only *Rosa* (rose) but also many edible fruit-bearing plants. *Malus* (apple), *Prunus avium* (cherry), *Prunus persica* (peach), *Prunus domestica* (plum), *Pyrus* (pear), *Rubus idaeus* (raspberry) and *Fragaria* (strawberry) (page 21) all belong to this family, and are economically very important. Successful pollination of these crops is vital so commercial bee-keepers are often called upon to bring their hives to the site to ensure that the highest rate of pollination is achieved. It has been claimed that fruit yields can be increased by anything up to 36% if bees are active on the crop, which is a huge advantage for growers. Other ornamental plants fall into this family too: for example, *Cotoneaster*, *Sorbus aucuparia* (rowan) and *Sorbus aria* (whitebeam).

Fabaceae

Like the 'rose' family, *Fabaceae* or (*Leguminoseae*) are economically extremely important. Legume seeds and foliage have a comparatively higher protein content than non-legume plants, a fact that has not gone unnoticed by mankind: we have taken advantage of this by growing edible legumes for ourselves, such as broad beans, runner beans, soybeans, chickpeas and peanuts, and also forage crops for animals, such as *Trifolium* (clover), *Vicia* (vetch), *Onobrychis viciifolia* (sainfoin) and *Medicago sativa* subsp. *sativa* (alfalfa or lucerne). In addition, because all legumes have the ability to fix atmospheric nitrogen they can be used in a crop rotation scheme to replenish soil that has been depleted of nitrogen, thereby reducing the need to use expensive artificial fertilizers. The *Fabaceae* family also includes ornamental plants such as *Lathyrus odoratus* (sweet pea), *Ulex europaeus* (gorse) and *Robinia pseudoacacia* (false acacia). It is worth noting, however, that some plants in this family – garden peas, for example – are self-fertile.

Brassicaceae

The name of this family is derived from one of its members, the *Brassica* genus, which is otherwise known as the cabbage family. It includes such edibles as cabbage, cauliflower, turnip and radish, as well as ornamentals like *Cheiranthus cheiri* (wallflower) (page 77) and *Iberis* (candytuft). From a commercial bee-keeper's point of view, however, by far the most important member of the family is *Brassica napus* subsp. *oleifera* (oilseed rape). Many people probably view its bright yellow flowers as a blot on the landscape and hay fever sufferers groan at the first whiff of its perfume (although according to the *British Medical Journal* there is no clear evidence that oilseed rape has any adverse effects on health). Love it or hate it, the crop now accounts for 11%, in terms of acreage, of all cultivated crops in the UK.

Apiaceae

A number of bee-friendly plants belong to the *Apiaceae* (or *Umbelliferae*) family, including some herbs and vegetables: carrot, parsnip, parsley, dill and fennel. If they are left to flower, you will notice the similarity between them, and immediately see why they are popular with bees: the flowers grow in umbels, or clusters, which form a flat-topped umbrella shape (hence the alternative family name of *Umbelliferae*), ideal for insects to land on. Less obvious family members are *Eryngium* (sea holly) (page 94) and *Astrantia* (masterwort).

Boraginaceae

This family contains over 2,000 species, some of which are very useful bee plants, including, not surprisingly, *Borago officinalis* (borage) (page 24). Others include *Anchusa officinalis* (common bugloss or alkanet), *Symphytum* (comfrey), *Myosotis* (forget-me-not), *Heliotropium* (heliotrope) (page 146), *Brunnera macrophylla* (Siberian bugloss), and *Echium vulgare* (viper's bugloss). All have blue or purple flowers and hairy stems and leaves. Most of them are herbs and are used for dye, or medicinally. The most obvious one in the medicinal category is *Borago officinalis* (page 24), which is also known as starflower. Starflower oil contains the richest known sources of natural Gamma Linolenic Acid (GLA), which is reputed to have a number of benefits.

Dipsacus sativus (teasel)

Dipsacaceae

There are over 350 species in the *Dipsacaceae*, or teasel, family. The ones we are likely to come across in the UK are mostly known as wild flowers although many have been cultivated and have found their way into the garden. The most common wild ones, apart from *Dipsacus*, the teasel (page 140), are *Knautia*, *Scabiosa* and *Succisa* (all of which have the common name scabious). All have a dense head of flowers, surrounded by bracts. In the case of *Dipsacus sativus* (fuller's teasel), the dried bracts, which turn into sharp recurved spines, have been used in the wool industry to card, or comb the wool, teasing out the fibres ready for further processing. Teasel heads have since been replaced by metal carding devices.

You will find the name of each plant's family given in the gazetteer that follows. It is interesting to see which ones keep cropping up – and also to conjure with the names of some of the lesser-known ones.

Annuals and biennials

Spring

Cheiranthus cheiri (Wallflower) (page 77)

(*Brassicaceae*)

Good for	Nectar and pollen
Hardy biennial	
Height	30cm
Sow seeds	March to May
Flowers	Various, including crimson, yellow, reddish brown; March to May the year after sowing
Best position	Sun
Soil	Moisture retentive but well drained, fertile

Wallflowers are beloved by bees and were once known as 'bee flower'. In the 17th century Gervase Markham wrote: 'The Husbandman preserves it most in his Bee-garden, for it is wondrous sweet and affordeth much honey.'

Until recently wallflowers were very popular in spring bedding schemes with stately tulips emerging through the understated layer of scented wallflowers. Their place seems to have been usurped by winter-flowering pansies, but for bees, wallflowers are a much better option. They come in a beautiful range of colours, my favourites being 'Vulcan', a really deep red and the pale yellow 'Primrose Dame'. Wallflowers are members of the cabbage family, and as such they can be susceptible to 'club root', where the plants droop and turn yellow: if this happens you must dig them up and burn the plants immediately. It also means that you cannot grow any member of the cabbage family in the same soil for a number of years.

———————— *Summer/autumn* ————————

Eschscholzia californica (Californian poppy)
(*Papaveraceae*)

Good for	Nectar and pollen
Hardy	
Height	30cm
Sow seeds	March or April (or September to flower following year) *in situ*
Flowers	Shades of orange and yellow, white; June to August
Best position	Sun
Soil	Moderately fertile, moist but well drained

This is a fast-growing annual that will flower in 8–10 weeks after a spring sowing. Thereafter it will happily self-seed and you may even find that it becomes invasive. To avoid this, and to ensure a long flowering period, keep deadheading. It has 'papery' flowers in shades of orange, yellow and white and is a lovely flower for cutting.

Iberis umbellata (Candytuft)
(*Brassicaceae*)

Good for	Nectar and pollen
Hardy	
Height	30cm
Sow seeds	March or April (or September to flower following year) *in situ*
Flowers	Shades of lilac, purple, pink, and red, white; June to August
Best position	Sun
Soil	Moderately fertile, moist but well drained

This is a lovely, sweet-scented annual which more often than not finds itself in the children's section of the seed display because it is so easy to grow. Try it as a 'gap-filler' in the herbaceous border, or sow some especially for cutting.

Callistephus chinensis (China aster)

(*Asteraceae*)	
Good for	Nectar and pollen
Half-hardy	
Height	50cm
Sow seeds	March/April under cover to be planted out later; May *in situ*
Flowers	Shades of purple, pink, and red, white; July to September
Best position	Sun
Soil	Well drained

Be careful which seeds you buy – there are many double varieties but fewer single ones, which are the ones we want in order to attract bees. Like many other annual flowers they are ideal for cutting. Try to grow China asters in a different place each year because they are susceptible to a number of soilborne diseases which can build up.

Clarkia elegans

(*Onagraceae*)	
Good for	Nectar and pollen
Hardy	
Height	40cm
Sow seeds	March or April (or September to flower following year) *in situ*
Flowers	Shades of purple and lilac, pink and white; June to October
Best position	Sun
Soil	Moisture retentive but well drained

Clarkia produces many flowers on one stem which, like so many other annuals, look good in a vase as well as in the garden. *Clarkia pulchella* 'Passion of Purple' is a particularly fine plant which has deeply cut lilac-purple flowers.

Centaurea cyanus (Cornflower) (page 39)
(*Asteraceae*)

Good for	Nectar and pollen
Hardy	
Height	90cm
Sow seeds	March or September/October *in situ*
Flowers	Blue, pink, lilac; June to September
Best position	Sun
Soil	Any as long as it is well drained

Centaurea cyanus is the well-known wild flower, once common in cornfields: John Gerard, writing in his herbal of 1597, notes that it grows 'among Wheat, Rie, Barley, and other graine' [*sic*]. Improved methods of seed cleaning mean that it is now rarely seen on agricultural land unless it has been sown as part of a conservation scheme. It does, however, make a splendid addition to the garden, especially as a cut flower. New varieties have extended the colour range to include pink, lilac, white and even maroon, but to my mind the original blue is still the best.

Cosmos bipinnatus (page 145)
(*Asteraceae*)

Good for	Nectar
Half-hardy	
Height	90cm
Sow seeds	March/April under cover to be planted out later; May *in situ*
Flowers	Shades of pink, white; June to October
Best position	Sun
Soil	Moisture retentive but well drained

Cosmos has single flowers held above feathery foliage – its simplicity makes it all the more attractive. There are a number of colours to choose from but my favourite is 'Purity', a white form. Newer introductions include 'Antiquity', which opens a rich burgundy but changes to bronze–salmon as it ages – almost like a faded tapestry. Again, it makes a beautiful cut flower.

***Heliotropium* arborescens** (Heliotrope or cherry pie plant) (page 146)
(*Boraginaceae*)

Nectar	
Half-hardy	
Height	45cm
Sow seeds	March/April under cover to be planted out later
Flowers	Violet; June to September
Best position	Sun
Soil	Moisture retentive but well drained

Some heliotrope are not strongly scented so, if you want to be sure to have some 'cherry pie' fragrance in your garden, grow the varieties 'Chatsworth' or 'Marine'. Heliotrope is not the easiest of plants to grow from seed: germination can be erratic so always sow more than you need. It can be propagated by cuttings taken during the summer but, not being fully hardy, it needs to be nurtured in a heated greenhouse over winter. Heliotrope looks spectacular as summer bedding but it also looks good grown in containers.

Cosmos bipinnatus

Nigella damascena, *N. hispanica* (Love-in-a-mist)
(*Ranunculaceae*)

Good for	Nectar and pollen
Hardy	
Height	45cm
Sow seeds	March to May; or September *in situ*
Flowers	Blue; July to September
Best position	Sun
Soil	Moisture retentive but well drained soil

The most common love-in-a-mist is *Nigella damascena*, and for many years 'Miss Jekyll' was the only variety available. Nowadays there are many more varieties, and colours, to be had but 'Miss Jekyll' is still very popular. My favourite, however, is *N. hispanica*, a distinct species from Spain which has deeply divided leaves and large, deep blue flowers which are much larger than those of *N. damascena*. Either species makes a good cut flower and, if you leave the flowers to form seed heads, these, too, look lovely in an arrangement.

Heliotropium arborescens (cherry pie plant)

Reseda odorata (Sweet mignonette)
(*Resedaceae*)

Good for	Nectar and pollen
Hardy	
Height	60cm
Sow seeds	March to May, or September *in situ*
Flowers	Green; July to September
Best position	Sun, partial shade
Soil	Moisture retentive but well drained

Sweet mignonette used to be seen in almost every garden – not because of its showy flowers but because of its scent (the name *odorata* is the giveaway) and the fact that it makes a wonderful cut flower, lasting many weeks and exuding its sweet scent throughout. It is related to wild mignonette (*Reseda lutea*) which is its wild, unscented counterpart.

Limnanthes douglasii (poached egg plant)

Limnanthes douglasii (Poached egg plant) (page 147)
(*Limnanthaceae*)

Good for	Nectar and pollen
Hardy	
Height	15cm
Sow seeds	April to July, or September *in situ*
Flowers	White with yellow centre; May onwards, from seeds sown in September, otherwise June to September
Best position	Sun
Soil	Moisture retentive but well drained

Looking at *Limnanthes* it is easy to see how it got its common name of poached egg plant! It is easy to grow, will self-seed readily, and looks a treat at the front of a border. It is also invaluable in the vegetable garden because as well as attracting bees, it is also popular with hoverflies which feed on aphids.

Helianthus annuus (Sunflower) (page 98)
(*Asteraceae*)

Good for	Nectar and pollen
Hardy	
Height	60cm–3m, depending on variety
Sow seeds	April onwards
Flowers	Yellow, amber, brown, orange and wine red; July to September
Best position	Sun
Soil	Moisture retentive but well drained

I have found that people tend to fall into one of two camps when it comes to sunflowers – you either love them or loathe them (a bit like Marmite!). I love them (and Marmite), especially the newer, shorter varieties with smaller flowers that are excellent for cutting. One of my favourites is 'Lemon Sorbet', a pale yellow variety which has a branching habit so that you get more than one flower on each main stem. Another is 'Pastiche' which has a range of more muted colours including yellow, red and buff. If you want to grow a really tall, 'traditional' variety, then try 'Russian Giant'. One endearing quality of sunflowers is that they track the sun: they start facing east in the morning and by evening they are facing west. Not only are they good 'bee' plants but if you leave the seed heads on the plant they will provide a winter feast for the birds.

Amberboa moschata (Sweet sultan)

(*Asteraceae*)	
Good for	Nectar and pollen
Hardy	
Height	50cm
Sow seeds	March to May
Flowers	Various including pink, white, purple and yellow; July to October
Best position	Sun
Soil	Moisture retentive but well drained

Originating in Turkey and the Caucasus, sweet sultan was introduced into Britain some 350 years ago. It is highly scented and looks a little like a cross between an aster and a cornflower. It makes a beautiful cut flower, and is very attractive to bees and other insects.

Flowering times of selected annual and biennial flowers

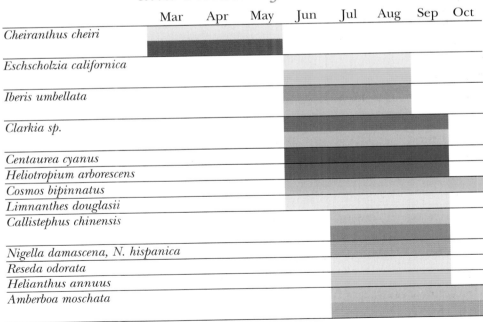

Figure 11 Flowering times of selected annual and biennial flowers

Herbaceous perennials and bulbs

———— Spring ————

Crocus

(*Iridaceae*)

Good for	Nectar and pollen
Hardy corm	
Height	10cm
Flowers	Funnel-shaped, long-tubed white, purple or yellow flowers; February to April
Best position	Sun or partial shade
Soil	Humus-rich, moist but well drained

There are two species of *Crocus* which provide bees with both nectar and pollen in early spring. Firstly, *C. chrysanthus* (or snow crocus), which contains more named cultivars than any other species. You will probably recognize some of them from your trip to the garden centre during the autumn: 'Cream Beauty', 'Blue Pearl' or 'Snow Bunting'. This species flowers from late winter to early spring.

Later comes *C. vernus*, or giant Dutch crocus. As its name implies, this *Crocus* has much larger corms and flowers than *C. chrysanthus*, which makes it more suitable for naturalizing. Varieties include 'Jeanne d'Arc' (white), 'Remembrance' (dark blue and purple with silvery gloss) and 'Yellow Mammoth', which speaks for itself!

Eranthis hyemalis (Winter aconite)

(*Ranunculaceae*)

Good for	Nectar and pollen
Hardy tuber	
Height	10cm
Flowers	Stalkless, cup-shaped yellow flowers; February to April
Best position	Partial shade
Soil	Humus-rich, moist but well drained

Because winter aconites like conditions that can ideally be found in damp woodland areas (their original habitat), and they flower in late winter and early spring, they are perfect for planting under deciduous shrubs and trees. They are sensitive to heat and will only open once the temperature reaches 10°C which, from our point of view, is ideal because this temperature is high enough for bees to be out foraging.

You can plant the tubers in autumn about 5cm deep, or you can buy plants 'in the green' in late spring, which are generally less expensive. If you buy the latter then plant them immediately you get them, to the depth at which they have already been growing – this will be obvious from the colour of the stems.

Do not confuse winter aconite with *Aconitum*, whose common name is also aconite – this is an herbaceous perennial which, although magnificent, is of little value to honeybees.

Anemone blanda (Windflower)

(*Ranunculaceae*)	
Good for	Pollen
Hardy tuber	
Height	5–10cm
Flowers	Flat blue, pink or white flowers; March and April
Best position	Sun or partial shade
Soil	Humus-rich, moist but well drained

Anemone is a large genus of perennial plants. One of the most common is the autumn-flowering herbaceous anemone *A.* x *hybrida*. There is also *A. coronaria* which produces rich, almost gaudy, coloured flowers from knobbly tubers from late spring. The one that we are interested in, however, is the spring flowering *A. blanda* which provides bees with a valuable source of pollen.

They are ideal for naturalizing so you should choose a spot in the garden where they can be left undisturbed.

Doronicum (Leopard's bane)

(*Asteraceae*)	
Good for	Nectar and pollen
Hardy	
Deciduous	
Height	Up to 60cm
Flowers	Yellow, daisy-like flowers; April to May
Best position	Sun or partial shade
Soil	Almost any except waterlogged

Doronicum is one of the earliest flowering perennials, bringing cheery daisy-like flowers to the border in spring. Varieties of *Doronicum orientale*, such as 'Magnificum', look lovely with tulips in the cultivated border, and the wild *Doronicum pardalianches* complements cowslips (*Primula veris*) beautifully. Its common name 'leopard's bane' comes from the fact that it is deadly to animals. In fact John Gerard, writing in his herbal of 1597 affirms that 'it killeth Panthers … Wolves, and all kinds of wilde beasts, being given them with flesh'. And it 'killeth dogs, it is very certain, and found out by trial'. And yet 'this herbe or the root thereof is not deadly to man' – I'm not sure I would like to put it to the test, though, despite what Gerard says!

Muscari (Grape hyacinth)
(*Hyacinthaceae*)

Good for	Nectar and pollen
Hardy bulb	
Height	15cm
Flowers	Small bell-shaped blue flowers on dense spikes; April to May
Best position	Sun
Soil	Well drained

The most common variety of grape hyacinth is *Muscari armeniacum*. It is a little gem of a plant, much loved by honeybees, which might at first glance seem odd because of the shape of the individual flowers. The 'bells' are shallow enough, and the nectar profuse enough, however, for the bees to access it. *Muscari* can be a little invasive because it spreads readily by seed so, if you do not want swathes taking over your garden, cut off the spent flower spikes before the seed ripens.

———— *Spring / summer* ————

Tulipa (Tulip) (pages 74–5)
(*Liliaceae*)

Good for	Pollen
Hardy bulb	
Height	Up to 45cm, depending on variety
Flowers	Goblet-shaped, upward-facing flowers in various colours; March to June, depending on variety
Best position	Sun
Soil	Well drained

There are so many different tulips, and their history is so intriguing, that whole volumes have been written about them, so whatever I say here will pale into insignificance! Suffice it to say that, although they do not come in the top ten of 'bee plants', they are a useful addition to the garden during the spring months to provide pollen at a time when colonies are starting to build up. Having said that, however, bees will still forage on plants over and above tulips if they are available.

Do not plant tulip bulbs too early in the autumn as they can suffer from a fungal disease called tulip fire which thrives in warmer weather. They are also best planted deep.

Lamium maculatum (Dead nettle) (page 211)

(*Lamiaceae*)	
Good for	Nectar and pollen
Hardy	
Semi-evergreen	
Height	20cm
Flowers	White or mauve-pink, depending on variety; March to July
Best position	Partial shade
Soil	Moist but well drained

Lamium maculatum is the cultivated, variegated species of the native dead nettle; its name *maculatum* is derived from the Latin *macula*, meaning spot or stain, which refers to the variegation on the stems and leaves.

As well as being a good 'bee plant' it is really useful as a ground cover in partially shaded areas. There are several good varieties ranging from 'White Nancy', which has white flowers and splashes of silver on the leaves, and 'Beacon Silver', which has silver foliage edged in green with magenta flowers, to 'Roseum', which has green foliage with a centre splash of white, and delicate, pale pink flowers.

Papaver orientale (Oriental poppy)
(*Papaveraceae*)

Good for	Pollen
Hardy	
Deciduous	
Height	Up to 1m
Flowers	Large, cup-shaped scarlet, pink, plum, pale lilac or white flowers; May and June
Best position	Sun
Soil	Well drained

Usually, when you say 'poppy' to anyone they immediately think of the red cornfield poppy (*Papaver rhoeas*) but the group known as oriental poppies is one of the stars of the herbaceous border, albeit briefly in May and June. What they lack in nectar they certainly make up for in pollen, which attracts no end of bees.

There are numerous varieties to choose from – in one nursery catalogue I counted more than 80! Here are three of my favourites. First, *Papaver orientale* 'Juliane': this is quite a short plant, reaching about 50cm, and I have rarely had to stake it. It has beautiful, pale pink flowers during May and June. Second, *P. orientale* 'Raspberry Queen': this grows up to 75cm and has the most amazing pink flowers, described by Bob Brown of Cotswold Garden Flowers as 'Barbara Cartland with running mascara'! Third, there is *P. orientale* 'Patty's Plum', whose flowers I tend to describe as 'watered-down blackberry juice', which does not do them justice. It looks lovely interspersed with *Allium* 'Purple Sensation'.

Achillea millefolium (Yarrow) (page 155)
(*Asteraceae*)

Good for	Nectar
Hardy	
Deciduous/Semi-evergreen	
Height	Up to 90cm
Flowers	Flat-topped 'platforms' of small individual cream, yellow, pink, pale lilac or cerise flowers; May to August, depending on variety
Best position	Sun or light shade
Soil	Well drained

There is a multitude of vernacular names for *Achillea*, the most common being yarrow, but it is also known as soldier's woundwort, staunchweed and bloodwort, names which point to its use as a medicinal herb to heal wounds and staunch the flow of blood. And, if you place some under your pillow at night and recite the following poem, you are supposed to dream of your future wife or husband:

> *Thou pretty herb of Venus' tree,*
> *Thy true name it is Yarrow;*
> *Now who my bosom friend must be,*
> *Pray tell thou me tomorrow.*

I cannot vouch for its reliability but stranger things have happened!

Achillea millefolium (yarrow)

Although the flower of the 'wild' *Achillea* is a rather nondescript chalky white colour the cultivated hybrids have some subtle and striking hues, although most varieties will fade in colour as the blooms mature. Three of the best, in my opinion, are: *Achillea millefolium* 'Lilac Beauty', which is pale lilac; *Achillea* 'Credo', a lovely soft yellow; and *Achillea* 'Pretty Belinda', which has deep pink flowers.

If you deadhead regularly then new flowers will be produced well into late summer. It is best to cut all the flowering stems back to ground level in mid-autumn, however, to allow the plants to build up before winter sets in.

Geranium (Cranesbill) (page 76)
(*Geraniaceae*)

Good for	Nectar
Hardy	
Deciduous, semi-evergreen	
Height	Up to 75cm, depending on species
Flowers	Shallow, cup-shaped white, pink, burgundy, lavender-blue or purple flowers; May to October, depending on species
Best position	Sun or partial shade
Soil	Almost any except waterlogged

There are over 400 species of geraniums (not to be confused with pelargoniums, the tender plants grown for bedding and pot plants), so you are spoilt for choice. Add to that the fact that they are easy to grow, come in a wide range of colours, have a long flowering period and they are brilliant bee plants, and you have the perfect plant for any garden.

To try and recommend even a dozen would be a difficult task, but I will draw a couple out of the lucky dip to give you an idea of their versatility. First, a tried and tested variety, *Geranium* 'Johnson's Blue': this variety has dark-veined, lavender-blue flowers from May to August and a slightly lax habit which makes it ideal as a ground cover. It has been awarded the Royal Horticultural Society's Award of Garden Merit (AGM).

Second, there is *Geranium phaeum* 'Margaret Wilson', which has creamy-white variegation on the leaves and lovely violet flowers from April to June. Its variegated leaves will shine out from a partially shady spot where it will happily grow.

Third, and quite different from the other two, is *Geranium* 'Dusky Crûg', which has large, chocolatey-maroon leaves and sugary-pink flowers which seem to go on for ever – I have seen one plant producing flowers from May right through to November. It is a small plant, rarely growing more than 25cm in height with a spread of 40cm, so is ideal for a small space. Grow it in the sun, too; its beautiful foliage will be all but invisible in the shade.

Centranthus ruber (Valerian)
(*Valerianaceae*)

Good for	Nectar and pollen
Hardy	
Deciduous	
Height	60cm/1m
Flowers	Heads of small, star-shaped red, pink or white flowers; May to October
Best position	Sun
Soil	Well drained

This perennial has a somewhat loose habit which makes it unpopular with some gardeners. Add to that the fact that it naturalizes very easily and you can understand why. I find the white version (*Centranthus ruber* 'Albus') a more acceptable garden plant but all of the colours, surprisingly even the red, are popular with bees.

—————————— *Summer* ——————————

Allium (Ornamental onion)
(*Liliaceae*)

Good for	Nectar and pollen
Hardy bulb	
Height	Up to 90cm
Flowers	Globes of individual star-shaped purple or white flowers; June and July
Best position	Sun
Soil	Well drained

Technically the name *Allium* refers to a whole host of bulbous plants of the onion family, but here we are looking at the ornamental ones which are grown for their flowers alone.

Alliums seem to have become the 'must-have' plant in any discerning gardener's patch these days, due in some part to its popularity at the Chelsea Flower Show and other prestigious horticultural events. They pop up there on a regular basis – or should I say, 'explode' there, because some do resemble fireworks in mid display. And, in case I'm beginning to sound a little disparaging, let me say that they are one of the 'show garden plants' that do transfer easily to a more modest venue like a back garden.

In my opinion, by far the best is *Allium hollandicum* 'Purple Sensation' (page 115), which has deep violet flower heads, the shade of which seems to complement every other colour. It looks stunning when planted between other herbaceous plants in the border, and can act as a unifying punctuation mark, giving a rhythmic coherence to a long border. There is also a white *Allium*, 'Mount Everest', which again is worth growing. The foliage can look a little untidy but it is nearly always camouflaged by the plants growing around it and so is not too much of an eyesore. Not all ornamental *Allium* are attractive to bees, however, but if you stick with these two, you should not be disappointed.

Anchusa azurea (Anchusa)
(*Boraginaceae*)

Good for	Nectar
Hardy	
Deciduous	
Height	Up to 1.2m
Flowers	Spikes of small, forget-me-not-like blue flowers; June and July
Best position	Sun
Soil	Well drained

Anchusa has one of the bluest flowers imaginable – almost on a par with gentians. It is fairly short-lived so will need to be replaced periodically. The best way to propagate it is by root cuttings taken in the winter. The best-known garden variety is 'Lodden Royalist', which holds the Royal Horticultural Society's Award of Garden Merit. A paler blue variety is 'Opal' and if you are looking for something a little shorter, then go for 'Little John', which grows to about 50cm. The taller varieties will undoubtedly need staking to stop them from flopping over.

There is a biennial version of Anchusa (*Anchusa capensis*) which is grown as an annual.

Stachys byzantina (Lamb's ears) (page 95)

(*Lamiaceae*)	
Good for	Nectar
Hardy	
Deciduous	
Height	Up to 40cm
Flowers	Spikes of small, tubular pink flowers; June and July
Best position	Sun
Soil	Well drained

The common name of 'lamb's ears' comes from the white, 'woolly' surface of the leaves, which are irresistible to touch. In fact this plant is often grown just for its leaves, but if you let it flower you will see that it belongs to the same family as mints and nettles, the *Lamiaceae*, which is beloved by bees. (Beware of buying *S. byzantina* 'Silver Carpet' if you want flowers; this variety rarely, if ever, produces them, so is of no use in a bee-friendly garden.) The garden plant is very closely related to its wild cousin, betony, *Stachys officinalis*.

Polemonium caeruleum (Jacob's ladder) (page 76)

(*Polemoniaceae*)	
Good for	Nectar and pollen
Hardy	
Deciduous	
Height	Up to 60cm
Flowers	Clusters of cup-shaped, lavender-blue flowers; June to August
Best position	Sun but can cope with partial shade
Soil	Moist but well drained

Jacob's ladder is a stalwart of the herbaceous border and is very easy to grow, although it is fairly short-lived and you may have to replace it every couple of years or so. The most common variety carries lavender-blue flowers, but there is a white-flowered variety (*Polemonium caeruleum* 'Album') and one with variegated leaves (*Polemonium caeruleum* 'Brise D'Anjou). The plant gets its common name from the arrangement of its leaves – it has pairs of opposite leaflets that look like a ladder. If you deadhead then you will get a second flush of flowers later in the season.

——————————————— *Summer/autumn* ———————————————

Eryngium (Sea holly) (page 94)

(*Apiaceae*)	
Good for	Nectar
Hardy	
Deciduous/evergreen, depending on variety	
Height	Up to 75cm
Flowers	'Squashed' globes of tiny blue metallic flowers surrounded by a ruff of spiky bracts; June to September
Best position	Sun
Soil	Well drained

There are dozens of different species of *Eryngium* – all are spectacular in their own way, but here are three which I think are particularly good. First, there is *Eryngium* x *tripartitum* 'Jade Frost', which has 'evergrey' pink-flushed leaves with cream edges and amethyst-coloured flowers. It is a very special variety, which is well worth growing, but may be difficult to get hold of unless you go to a specialist nursery. Second, *Eryngium bourgatii* 'Graham Stuart Thomas': this has grey and white striped foliage and violet-blue flowers. Third, there is *Eryngium* x *oliverianum*, which has amethyst-coloured stems and amazing vivid blue flowers and holds the Royal Horticultural Society's Award of Garden Merit.

All species are attractive to bees because each tiny flower on the head produces abundant nectar.

Nepeta (Catmint) (page 7)

(*Lamiaceae*)	
Good for	Nectar
Hardy	
Deciduous	
Height	Up to 90cm, depending on variety
Flowers	Short spikes of lavender-like flowers; June to September
Best position	Full sun but can cope with light shade
Soil	Well drained

The common name of this plant indicates that it is a magnet for cats – they will roll around on the plant, flattening it entirely if given the chance. It is worth pushing a few stout twigs into the centre of the plant, taking care not to damage the crown, to act as a deterrent. I have heard it said that if you raise the plant from seed sown directly into the ground then cats will ignore it, but I have found no evidence to support this.

One of the most commonly grown varieties is 'Six Hills Giant', which is perhaps the tallest of all. It can be a little ungainly, so a good alternative is *Nepeta racemosa* 'Walker's Low', which grows to about 50cm and has beautiful violet/lilac flowers and, like all catmints, grey-green foliage. If you cut back the flower spikes after the first flush in June it will throw up more flowers which will last into early autumn.

Verbena rigida (page 59), *V. bonariensis*

(*Verbenaceae*)	
Good for	Nectar
Half-hardy	
Deciduous	
Height	80cm/1.5m
Flowers	Flat heads of lavender-purple flowers; June/August to October
Best position	Sun
Soil	Well drained

Most people are familiar with *Verbena bonariensis*, which has become the 'must-have' plant of the past decade due to its use by prominent designers in show gardens at Chelsea and the like. There is no doubt it has a certain *je ne sais quoi*: its 'see-through' stems add a romantic touch to any border, and the purple flowers set off yellows and oranges to perfection. Equally lovely is the shorter *Verbena rigida*, which flowers earlier. Both species, however, are a little tender, struggling in a wet or particularly cold winter, so protect them with mulch in early winter or, to make doubly sure you have plants for next year, take cuttings in the autumn. They grow easily from seed, however, so if you do lose yours and you have forgotten to take those cuttings, all is not lost – just start again in the spring.

Both species are attractive to bees and other nectar-seeking insects.

Helenium (Sneezeweed) (page 81)
(*Asteraceae*)

Good for	Nectar and pollen
Hardy	
Deciduous	
Height	Up to 90cm
Flowers	Yellow, through orange to red, daisy flowers with prominent centre; June to October
Best position	Sun
Soil	Well drained

Sneezeweed comes from North America, where the Native Americans dried the leaves to make a snuff which promoted sneezing, to get rid of evil spirits inhabiting the body – hence the common name.

Helenium is usually seen in flower throughout autumn, but there are some earlier-flowering varieties which add a fillip to the border, such as the bronze-coloured *Helenium* 'Abbey Dore' and yellow *H*. 'El Dorado', both of which start to flower in June. Perhaps the best-known variety, however, is the rich red *H*. 'Moerheim Beauty', which flowers from July. Two of my favourites are *H*. 'Beatrice', which has tawny yellow flowers and *H*. 'Red Jewel', which has rich, dusky-red flowers from July to September.

As well as being a good bee plant, *Helenium* are lovely for cutting.

Gaillardia (Blanket flower) (below)

(*Asteraceae*)	
Good for	Nectar and pollen
Hardy	
Deciduous	
Height	Up to 60cm
Flowers	Daisy flowers in red, yellow, orange/red or yellow/orange; June to November
Best position	Does best in sun but can cope with light shade
Soil	Well drained

Gaillardia is named after an 18th-century French magistrate, M. Gaillard de Charentonneau, who was a patron of botany, and its common name (blanket flower) comes from its resemblance to the patterned blankets made by Native Americans in the plant's indigenous home of the prairie provinces of the North American continent.

Gaillardia x grandiflora 'Dazzler' (blanket flower)

This is probably one of the best 'daisy' flowers to plant in a 'hot' border – colours range from yellow through orange to red, and the flowers are not only very long-lasting but make ideal cut flowers. As if that were not enough, they are excellent bee plants too!

If you want a variety that really stands out from the crowd, then choose *Gaillardia* x *grandiflora* 'Dazzler', which has flowers with an orange/red centre tipped with yellow; it is been awarded the Royal Horticultural Society's Award of Garden Merit, so it is well worth growing. A good all-red variety is *Gaillardia* x *grandiflora* 'Burgundy', which has lovely dark red flowers with a yellow centre.

Echinops ritro, *E. bannaticus*, *E. sphaerocephalus* (Globe thistle) (page 94)
(*Asteraceae*)

Good for	Nectar
Hardy	
Deciduous	
Height	Up to 120cm, depending on species
Flowers	'Bristly', metallic-blue pompom flowers; July to September
Best position	Sun or partial shade
Soil	Well drained

This is a distinctive, architectural plant with spherical blue flowers (*E. sphaerocephalus* has off-white, pale lilac ones) held on stiff stems above spiky, grey-green foliage. It is beloved of bees and they will nearly always ignore other blooms in favour of *Echinops* when it is in flower.

Sedum spectabile (Ice plant) (page 81)
(*Crassulaceae*)

Good for	Nectar
Hardy	
Deciduous	
Height	Up to 60cm
Flowers	Flat heads of pink or white flowers; August and September
Best position	Sun
Soil	Well drained

Along with Michaelmas daisies, *Sedum* are valuable autumn-flowering plants to have in the border, providing an important source of nectar for butterflies as well as bees.

Heads of hundreds of tiny flowers are held above glaucous-blue, fleshy, succulent leaves which appear early in the growing season. There are a number of worthy varieties, including *Sedum spectabile* 'Brilliant', which has bright pink flowers, *S. spectabile* 'Stardust', with white flowers, and perhaps the most popular of all, *Sedum* 'Herbstfreude' or 'Autumn Joy', which has rose-red flowers turning to russet late in the season.

Kniphofia (Red hot poker) (page 166)

(*Asphodelaceae*)	
Good for	Nectar
Hardy	
Deciduous semi-evergreen with some protection	
Height	Up to 1.2m, depending on variety
Flowers	Spikes of orange, yellow and cream flowers; August to September
Best position	Full sun but can cope with light shade
Soil	Fertile, well drained

Neither its common name nor its botanical name endears it to the average gardener, but *Kniphofia* provides a really useful 'punctuation mark' in the herbaceous border, and it is an excellent bee plant. This may not be obvious, given that its flowers are tubular, but the plant produces so much nectar that you can actually see it seeping out of the flowers. Beware wasps, however – *Kniphofia* seems to attract them like no other herbaceous plant.

At one time the only *Kniphofia* that was available had orange/red flowers, resembling – yes, you've guessed – a red hot poker! Nowadays, however, lots of new varieties have been introduced which are more subtle and less garish than the original. I am thinking of varieties like 'Little Maid', which has flowers that start pale green, and then fade to cream. The leaves are narrow and neat, and it is a short variety, rarely growing more than 60cm tall. Two more of my favourites are *Kniphofia* 'Toffee Nosed', which has tawny orange flowers fading to cream, and *Kniphofia* 'Bee's Lemon' (what else!), which has green buds opening to lemon-yellow flowers.

Aster (Michaelmas daisy) (page 82)
(*Asteraceae*)

Good for	Nectar and pollen
Hardy	
Deciduous	
Height	Up to 1.5m, depending on variety
Flowers	Daisy flowers in purple, lavender-blue, pink or white; August to October
Best position	Sun or can cope with light shade
Soil	Well drained

The common name is indicative of the time of year that this group of *Aster* flowers, Michaelmas being celebrated on 29 September. It is one of the best autumn-flowering plants, although many gardeners avoid it because of its susceptibility to mildew. It is true that some older varieties suffer from this affliction, but newer ones appear to be less vulnerable. This is a minor drawback when you consider its value as a bee plant. The flowers provide quantities of nectar and especially pollen at a time of year when other sources are diminishing.

One of the best varieties is *Aster* x *frikartii* 'Mönch'. It has masses of lavender-blue flowers with orange centres held above dark green leaves. I also like *Aster novae-angliae* 'Andenken an Alma Pötschke', with cerise flowers, and *Aster novae-angliae* 'Violetta', with rich violet/purple flowers. If you're looking for a shorter variety then consider *Aster amellus* 'Veilchenkönigin', a deep violet blue, which grows to about 30cm and is ideal to fill a gap at the front of the border. There are some varieties with paler flowers, but I think they look a little washed out in the fading autumn light, especially if they are planted alongside the stronger hues characteristic of many autumn-flowering plants.

Kniphofia sp. (red hot poker) (opposite)

Rudbeckia (Coneflower or black-eyed Susan) (page 31)
(*Asteraceae*)

Good for	Nectar
Hardy	
Deciduous	
Height	60cm
Flowers	Yellow, through orange to red, daisy flowers with dark centre; August to October or first frosts
Best position	Sun or partial shade
Soil	Well drained

Like *Helenium, Rudbeckia* are natives of North America, have daisy-like flowers, and are very good for cutting. They are also excellent bee plants.

There are a number of species and varieties that are suitable for the garden, including *Rudbeckia laciniata* 'Herbstsonne', which has bright yellow flowers, and *Rudbeckia fulgida* var. *sullivantii* 'Goldsturm', again with yellow flowers. If you fancy something new, try *R.* x *hirta* 'Cherry Brandy', which is a recent introduction and the first red variety grown from seed.

Solidago (Golden rod)
(*Asteraceae*)

Good for	Nectar and pollen
Hardy	
Deciduous	
Height	Up to 2m, depending on variety
Flowers	Yellow panicles; August to October
Best position	Sun or partial shade
Soil	Well drained

Golden rod is one of those flowers you either love or hate. Most people I have spoken to seem to hate it, usually because they have grown it in their garden and it has taken over, crowding out and bullying less robust plants. If you have room, it is worth growing, especially *Solidago rigida*. This looks a little like ragwort but bees love it. It flowers from August to October and grows to about 90cm. Also worth a mention is *Solidago* 'Laurin', which has butter-yellow, featherlike stems of flowers and only grows to about 60cm.

Perovskia atriplicifolia (Russian sage) (page 89)

(*Lamiaceae*)

Good for	Nectar
Hardy	
Deciduous	
Height	Up to 1.2m
Flowers	Violet/blue; August/September
Best position	Sun
Soil	Well drained, poor to moderately fertile – can cope with chalky soils and coastal conditions

Russian sage is technically a subshrub (a plant that is woody at the base but whose terminal shoots die back in winter) but is often classified as a herbaceous perennial because that's how most people use it. It has aromatic, silver-grey leaves and spikes of violet-blue flowers, not dissimilar to lavender. It can grow to quite a size, but there is a cultivar, *Perovskia atriplicifolia* 'Little Spire', which only grows to 60cm, making it a better option for a smaller border. Despite its common name of Russian sage, it is not edible.

Liatris spicata

(*Asteraceae*)

Good for	Nectar
Hardy	
Deciduous	
Height	60cm
Flowers	Dense spikes of small, purplish-pink or white flowers; August and September
Best position	Sun
Soil	Fertile, moist but well drained

These spikes of flowers make an attractive addition to the border, providing useful 'punctuation marks' in the midst of other late summer- and autumn-flowering perennials. If the usual purplish pink is a little too garish for your design then opt for the white version (*L. spicata* 'Alba'). They are also ideal for cutting.

Dahlia – single-flowered varieties only (page 111)
(*Asteraceae*)

Good for	Nectar and pollen
Tender tuber	
Height	Up to 1.5m, depending on variety
Flowers	Open white, pink, yellow, orange or red flowers; August to October/first frosts
Best position	Sun
Soil	Well drained

I'm not sure why, but when I ask people whether or not they grow dahlias, the reply always seems to be, 'No, but my granddad did'! Are dahlias 'ageist'? Do you have to be of a certain age – and gender – before you are allowed to grow them? I hope not; in fact I think dahlias are undergoing a bit of a renaissance, especially, I'm pleased to say, the single-flowered varieties which are just right for our buzzy friends. They (the single dahlias) seem to blend in with other plants so much more readily than the more flamboyant and fancier double varieties, adding extra colour and form to the late-summer autumn border. They also make excellent cut flowers.

There are so many dahlias to choose from, but three of my favourites are: 'Dark Desire', which has chocolatey-red single flowers with a yellow centre (the bit that the bee sees!); 'Honka', which has pale yellow, 'spidery' flowers; and 'Magenta Star', which has purple-pink flowers above dark foliage – lovely!

Dahlias are tender, however, and need a little more care and attention than hardy perennials. Do not plant out cutting-raised plants or dry tubers until all risk of frost has gone. When the first frost of the autumn blackens the foliage, lift the tubers, dry them off and then pack them in bone-dry compost in a frost-free place over winter.

Echinacea (Coneflower) (page 82)

(*Asteraceae*)	
Good for	Nectar
Hardy	
Deciduous	
Height	60cm
Flowers	Pink/purple, white/cream daisy flowers with dark centre; August to October or first frosts
Best position	Sun or partial shade
Soil	Moist but well drained

This is where using the common name for a plant can cause some confusion, since both *Echinacea* and *Rudbeckia* are known as coneflower, and I have even seen *Echinacea* listed as *Rudbeckia* in a nursery catalogue.

There are two ways of telling *Echinacea* and *Rudbeckia* apart. First, there is the colour. *Echinacea* are usually in the pink/purple colour range – indeed the most common *Echinacea* is *E. purpurea*, which is a giveaway! You will occasionally find white or creamy/yellow varieties, such as *E.* 'White Swan' or *E.* 'Sunrise' respectively, which can be a little confusing, since nearly all *Rudbeckia* are in the yellow/orange/gold part of the colour wheel. (There are exceptions, of course, since you can now find mahogany/red varieties of *Rudbeckia*.)

The simplest way of telling them apart, however, is to look at the centre of the flower. *Echinacea* comes from the Greek *echinos* which, roughly translated, means 'spiny'. The centre of *Echinacea* does indeed feel slightly prickly or spiny whereas the centre of *Rudbeckia* is much softer.

Whichever you choose, *Echinacea* or *Rudbeckia*, you can be sure to be offering the bees some food.

Autumn

Colchicum autumnale (Autumn crocus or naked ladies) (page 112)

(*Colchicaceae*)	
Good for	Nectar and pollen
Hardy corm	
Height	15cm
Flowers	Goblet-shaped pink/purple flowers; September and October
Best position	Sun or partial shade
Soil	Humus-rich, moist but well drained

Colchicum are often known as autumn crocus, which is confusing because they are not related to true crocuses at all, although they do flower in autumn. Their other common name, 'naked ladies', is perhaps more apposite, because the flowers emerge from the earth without any leaves, which appear much later. Beware, though – all parts of the plant are poisonous.

Flowering times of selected perennial flowers and bulbs

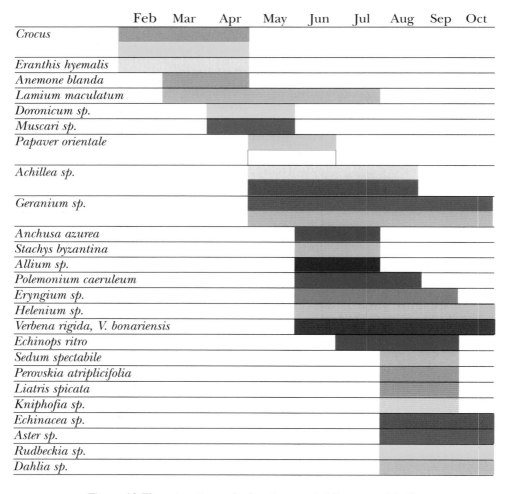

Figure 12 Flowering times of selected perennial flowers and bulbs

Shrubs and climbers

— All year round —

Ulex europaeus (Gorse)

(Fabaceae)

Good for	Nectar and pollen
Hardy	
Evergreen	
Height	Up to 2m
Flowers	Yellow; all year round
Best position	Sun
Soil	Any, even very poor

Gorse may not be the first 'must-have' shrub that springs to a gardener's mind – in fact many would dismiss it as not being gardenworthy at all – but it would certainly be in a bee's top ten shrubs. The main reason for this is that it has the capability to produce flowers all the year round, and generates prolific amounts of pollen. Although the common gorse can grow up to 2 metres high, *Ulex minor* and *Ulex gallii* form lower-growing shrubs, up to about 60cm, that the reticent gardener may be willing to find a home for.

Erica (Bell heather) (page 93)

(Ericaceae)

Good for	Nectar and pollen
Hardy	
Evergreen	
Height	Up to 60cm
Flowers	Varies throughout the year; depending on species
Best position	Sun
Soil	Varies, depending on species

There are hundreds of different varieties of *Erica*. Some flower during the winter, others during spring – in fact you can choose plants so that you have at least one in flower at any one time throughout the year. It is *Erica cinerea*, however, that is the native heather and is often seen growing alongside *Calluna* on heaths and moors. It flowers from July to September and needs an acid soil to thrive. Here are three recommendations of cultivated varieties for the garden: 'C.D. Eason' has bright magenta flowers with dark green leaves and is ideal for ground cover; 'Pink Ice' has rose-pink flowers, dark green foliage and a dwarf, twiggy habit; 'Godrevy' has white flowers and is compact and slow growing. Even if you do not have the correct soil in your garden to grow heathers successfully, it is worth planting some in a container, where you can use ericaceous compost, just to see the bees flocking to it to gather nectar.

Winter/spring

Viburnum (*V.* x *bodnantense* and *V. tinus* (page 73))
(*Caprifoliaceae*)

Good for	Nectar and pollen
Hardy	
Height	Up to 3m
Flowers	*V.* x *bodnantense* – pink; November to March
	V. tinus – white; December to May
Best position	Sun but can cope with partial shade
Soil	Well drained

Although you are unlikely to see many bees out and about during the deepest winter months, the occasional bee (especially bumble) might make an early spring foray and these two viburnum will provide both nectar and pollen for them when there are few other sources available.

In my opinion the best *Viburnum* x *bodnantense* to grow is 'Dawn': it has clusters of pink flowers which appear before the leaves. If you have one in your garden then its sweet fragrance will permeate the air for metres around. In a cold location the flowers may not appear until January, but in warmer situations they can be seen as early as November.

'Eve Price' is the best *Viburnum tinus*: it only grows to about 2 metres, is evergreen and has tiny star-shaped white flowers on flattened heads which appear in late winter. Both have been awarded the Royal Horticultural Society's Award of Garden Merit.

Daphne (*D. mezereum*)

(*Thymelaeaceae*)	
Good for	Nectar and pollen
Hardy	
Height	1m
Flowers	Pink; February to March
Best position	Sun or partial shade
Soil	Well drained

Along with *Viburnum*, *Daphne mezereum* is a useful early-flowering shrub to have in the garden for nectar and pollen. The rosy-pink flowers are produced before the grey-green leaves appear, but if you find the pink a little too rosy, there is a white variety, 'Alba'.

There are numerous other *Daphne* worth growing in the garden, particularly the early-summer-flowering *Daphne* x *burkwoodii*, although by that time there are many other plants providing excellent sources of both pollen and nectar. Nevertheless, it is worth making room for some flowering shrubs, which give structure and 'backbone' to any border.

———————————— *Spring* ————————————

Mahonia aquifolium (Oregon grape)

(*Berberidaceae*)	
Good for	Nectar and pollen
Hardy	
Evergreen	
Height	Up to 1.5m but can be pruned hard
Flowers	Scented, deep-primrose clusters; March to May
Best position	Shade or semi-shade, but will tolerate sun
Soil	Moderately fertile, moist but well drained

This really is one of the most useful shrubs to have in your garden for bees. It has a profusion of flowers at a time when there is often little else available, and blue-black berries in the autumn, which birds adore. In addition, the dark green, holly-like leaves provide attractive autumn colour and winter structure.

Prunus laurocerasus **'Otto Luyken'** (Cherry laurel) (below)
(*Rosaceae*)

Good for	Nectar and pollen
Hardy	
Evergreen	
Height	Up to 1.25m but can be pruned
Flowers	White; April
Best position	Sun or semi-shade
Soil	Any except waterlogged

There are some cherry laurels which can reach 6 metres, which is fine if you have space. For my money, though, the cultivar 'Otto Luyken' is the best one to grow since it is of smaller stature and, if it still gets too big, it can be pruned by cutting hard back after flowering.

Prunus laurocerasus 'Otto Luyken' (cherry laurel)

Berberis (Barberry)

(*Berberidaceae*)	
Good for	Nectar and pollen
Hardy	
Deciduous or evergreen, depending on variety	
Height	Up to 3m but some varieties much less
Flowers	Variable – yellow, orange or red depending on variety; mainly April/May
Best position	Sun or semi-shade
Soil	Any except waterlogged

There are numerous varieties of *Berberis*: some are evergreen, others are deciduous; some are tall and sprawling, others are small and compact. What they all have in common, however, is their ability to attract bees – and their vicious thorns which clothe the branches. One of the best newcomers is the evergreen *Berberis darwinii* 'Compacta', which grows up to a metre high and has a very dense habit, making it ideal for the smaller garden. It is also ideal to use as a hedge, forming an almost impenetrable barrier. It has rich orange flowers followed by blue berries in the autumn, and small, glossy leaves which, when new, are tinged with red.

———————— *Spring/summer* ————————

Cotoneaster

(*Rosaceae*)	
Good for	Nectar and pollen
Hardy	
Deciduous or evergreen, depending on variety	
Height	Up to 4m, depending on variety
Flowers	White/pink; May/June
Best position	Any
Soil	Any

Cotoneasters are so ubiquitous in the British garden that you barely give them a second glance. But to bees they are one of the most attractive plants in the garden when they are in flower. There are numerous varieties but one of the most popular is *Cotoneaster horizontalis*. As its name implies, its normal growth is along the ground, but if you plant it against a wall you will appreciate its herringbone pattern of branches when the leaves fall in the autumn. It rarely grows more than a metre, but at the other end of the scale is *Cotoneaster lacteus*, which can grow up to 4m. *C. lacteus* is an evergreen with large, dark green leaves held on arching branches. It holds its fruit well into the winter and I have used some branches as decoration in the house at Christmas when I have not been able to get hold of holly.

Clematis (*C. armandii, C. montana, C.* x *jackmanii* and *C. viticella*) (*Ranunculaceae*)

Good for	Nectar and pollen
Hardy	
Height	*C. armandii* – up to 5m
	C. montana – up to 9m
	C. x *jackmanii* – up to 4.5m
	C. viticella – up to 6m
Flowers	*C. armandii* – March to May
	C. montana – May
	C. x *jackmanii* – July to September
	C. viticella – July to September/October
Best position	Sun or partial shade
Soil	Well drained

Although clematis provide little in the way of nectar, they are valuable pollen producers with species in flower from March right through to October. The first to flower is *C. armandii*, an evergreen plant, with dark, leathery leaves and fragrant white flowers. Some people have reported it as being on the tender side, but I have had one on a north-facing wall, in quite a breezy situation, and have had to be fairly ruthless in pruning it back because it has grown so well. Be careful if you grow an early-flowering *Viburnum* or *Daphne*, though, because the fragrances can sometimes 'clash' with one another.

Next in the flowering calendar comes *C. montana*, of which there are dozens of varieties in varying colours, but the 'original' is white. It is a vigorous climber which needs a stout support, such as a pergola or wall, or better still, let it climb into a tree. This clematis needs no pruning, except to cut out dead or damaged growth after flowering.

Then there is *C.* x *jackmanii*. Again, there are numerous varieties to choose from, but perhaps the most popular is 'Superba', which has large, velvety, dark purple flowers. Because it is a late-flowering clematis pruning is required although this is straightforward enough – just cut back the stems to a pair of strong buds about 20cm above ground level in February or early March, before new growth starts to appear.

Finally we have the *viticella* group. These clematis are the ones to grow if you have trouble with clematis wilt (a fungus which enters the plant through a wound and quickly causes rapid wilting and can kill the plant). A lovely variety is *C. viticella* 'Alba Luxurians', which has white and green flowers with a hint of pale violet. Another, less intricately coloured, one is *C. viticella* 'Etoile Violette', which has velvety, deep purple flowers. As with other late-flowering clematis, pruning should be carried out in early spring by simply cutting back all the old growth to a pair of strong buds about 20cm above ground level.

Summer

Rosa (Rose)

(*Rosaceae*)

(*Rosa forrestiana* (rosy crimson), *R. macrophylla* (clear pink), *R. moyesii* (dark pink), *R. rugosa** (rose) (page 91) and *R. rugosa alba** (white))

Good for	Pollen
Hardy	
Deciduous	
Height	1.8m – 3m, depending on variety
Flowers	Single flush in June except those marked *, which are repeat flowering
Best position	Sun or partial shade, sheltered
Soil	Does best in fertile, well-drained but moisture-retentive soil, but can tolerate poorer, even clay, soil

According to a survey done for the television programme *Gardener's World* the rose is the nation's favourite flower. I suspect, however, that when most people think of a rose it is the quintessentially English double bloom that comes to mind and not the single blossom of the species rose. This is unfortunate, because it is the latter to which bees are most attracted. Some of the best 'bee' roses are listed above and, although they are not as flamboyant as the double varieties, they have a charm all of their own, and they produce beautifully coloured hips in the autumn.

Philadelphus (Mock orange)
(*Hydrangeaceae*)

Good for	Nectar and pollen
Hardy	
Deciduous	
Height	Up to 4.5m depending on variety, but can be hard pruned
Flowers	White; June/July
Best position	Sun or partial shade
Soil	Moist but well drained

Mock orange has a distinct orange blossom fragrance (hence its common name) and is extremely attractive to bees. There are over 75 species and cultivars of *Philadelphus*, so there is bound to be one to suit your garden, but be sure to choose one that has single flowers. *Philadelphus microphyllus* is a fairly low-growing variety (about a metre) and has small, glossy leaves and, of course, single flowers. If you want a larger shrub then try *Philadelphus* 'Beauclerk': this will grow up to 2 metres with large single flowers which are a good 5cm in diameter.

Leptospermum scoparium (Manuka or tea tree)
(*Myrtaceae*)

Good for	Nectar and pollen
Half-hardy	
Evergreen	
Height	Up to 3m
Flowers	White; June to August
Best position	Full sun, sheltered
Soil	Fertile, well drained

This is the shrub, which grows wild in New Zealand, from which the renowned manuka honey comes from. Manuka honey commands a high premium because of its reported medicinal benefits, although all honey (especially heather honey) has valuable antiseptic properties. Although it has another common name of 'tea tree' this is not the shrub that tea tree oil comes from, which is *Melaleuca alternifolia* – a good example of where knowing the Latin name is of benefit.

Leptospermum will thrive in milder, warmer parts of the UK but elsewhere it needs some tender loving care, especially over the winter months when more than a couple of degrees of frost will sound a death knell. Despite the fact that it needs a little more looking after, it is well worth growing because bees love it.

Hydrangea aspera 'Macrophylla' (page 9)

(*Hydrangeaceae*)	
Good for	Nectar and pollen
Half-hardy – may need some protection in winter	
Deciduous	
Height	Up to 2m
Flowers	Purple/white; July/August
Best position	Sun or partial shade
Soil	Does best in fertile, well-drained but moisture-retentive neutral or slightly acid soil, but can tolerate poorer soil

Hydrangeas are popular garden plants, but not all hydrangeas are suitable for bees. The large 'mop head' varieties are sterile or semi-sterile and of no use to bees. It is the 'lace cap' varieties which are the ones to grow, and in particular I have singled out *Hydrangea aspera* 'Macrophylla'. It is a handsome shrub with a mass of small, fertile flowers, surrounded by larger, sterile flowers. On a visit to Levens Hall in Cumbria towards the end of August, I was mesmerized by one such shrub positively alive with honeybees – a real treat for both gardener and bee-keeper.

——————— *Summer/autumn* ———————

Eucryphia x *nymansensis* 'Nymansay' (Leatherwood) (page 109)

(*Eucryphiaceae*)	
Good for	Nectar and pollen
Mostly hardy	
Evergreen	
Height	Up to 10m (4m in ten years)
Flowers	White, open cup-shaped, large flowers; August and September
Best position	Sheltered in full sun
Soil	Well drained, moisture retentive, fertile

This is an invaluable shrub for bees during late summer and early autumn – they flock to it like, well, bees to a honey pot! You do need space for this shrub because, given suitable conditions, it can grow up to ten metres and it seems a shame to prune or cut it back, with the inevitable loss of flowering material, just to keep it within bounds.

Calluna vulgaris (Ling heather) (page 93)

(*Ericaceae*)	
Good for	Nectar and pollen
Hardy	
Evergreen	
Height	45cm
Flowers	Varies; August to November, depending on variety
Best position	Sun
Soil	Acid

The name *Calluna* derives from the Greek *kallunein*, meaning to cleanse; so the name could come from the fact that heather twigs were used as brooms, or from the plant's medicinal properties.

You can see purple cushions of ling heather clothing heaths and moorlands throughout Britain in August and September – it really is a sight to behold. For the garden, though, there are hundreds of cultivated varieties which are worth growing and which will attract no end of bees – here are but three: *C. vulgaris* 'Anthony Davis' has grey leaves and white flowers; 'Joy Vanstone' has gold leaves which turn bronze in autumn, and mauve-pink flowers; 'Wickwar Flame' has gold leaves which turn orange and then red, and lilac flowers. All hold the Award of Garden Merit from the Royal Horticultural Society.

It is ling heather which produces the 'real' heather honey.

Flowering times of selected shrubs and climbers

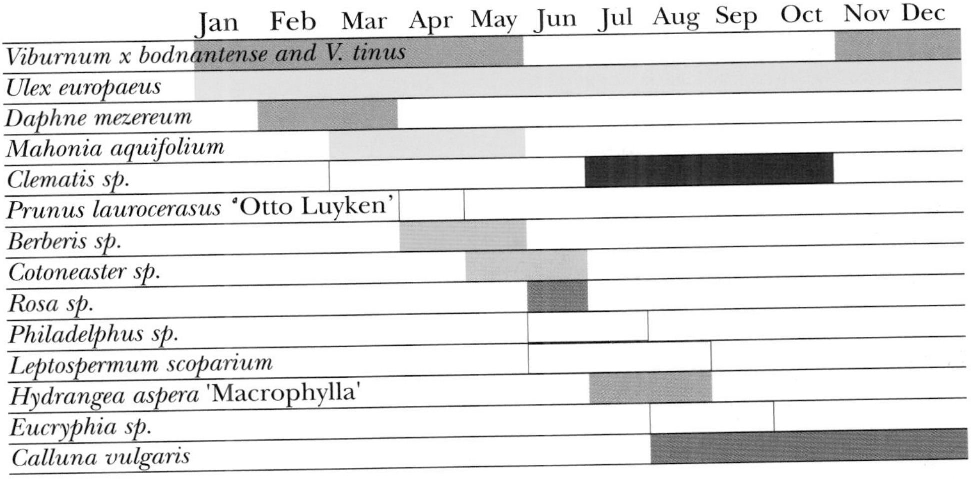

	Jan	Feb	Mar	Apr	May	Jun	Jul	Aug	Sep	Oct	Nov	Dec
Viburnum x bodnantense and V. tinus												
Ulex europaeus												
Daphne mezereum												
Mahonia aquifolium												
Clematis sp.												
Prunus laurocerasus 'Otto Luyken'												
Berberis sp.												
Cotoneaster sp.												
Rosa sp.												
Philadelphus sp.												
Leptospermum scoparium												
Hydrangea aspera 'Macrophylla'												
Eucryphia sp.												
Calluna vulgaris												

Figure 13 Flowering times of selected shrubs

Trees

Winter/spring

Corylus avellana (Hazel) (page 184)

(*Betulaceae*)	
Good for	Pollen
Hardy	
Deciduous	
Height	Up to 10m
Flowers	Catkins appear before leaves; February to April
Best position	Sun or part shade
Soil	Moderately fertile, moist but well drained

The common hazel has yellow pollen-bearing catkins (sometimes known as lamb's tails) from February to April, which are a valuable early source of pollen for bees. It also produces edible nuts in the autumn.

Technically the hazel is a tree, but it can be coppiced to produce straight branches which are used to make hurdles, beanpoles and stakes for hedge laying. Coppicing is where the plant is cut back to low stumps (or stools) every few years to promote the growth of straight stems. If the tree is coppiced it seldom grows higher than 6 metres.

Corylus avellana (hazel)

Spring

Salix caprea (Goat willow)

(*Salicaceae*)	
Good for	Pollen
Hardy	
Deciduous	
Height	6–9m
Flowers	Catkins; March to April
Best position	Warm, sunny site
Soil	Moderately fertile, moist but well drained

Goat willow is a large shrub or small tree that bears catkins at the end of its branches in the early spring. It is the male form, with ovoid catkins rather like powder puffs, which is attractive to bees. The female form has narrower, longer catkins which are grey (the well-known 'pussy willow'). In order to make sure you buy the correct form, wait until flowering time and then buy a containerized plant.

As well as being useful to bees the goat willow has also served man well – all willow bark contains salicin, from which aspirin was originally derived, and the stems are very flexible, making them ideal for use in basket making.

Aesculus hippocastanum (Horse chestnut)

(*Hippocastanaceae*)	
Good for	Nectar and pollen
Hardy	
Deciduous	
Height	Up to 24m!
Flowers	White with small yellow spots which turn pink; April/May
Best position	Full sun or partial shade
Soil	Any except waterlogged

This is also known as the conker tree. Conkers (or 'cheesers', 'cheggies' or 'obblyonkers', depending on where you live), beloved of schoolchildren throughout the country, are the seeds of the horse chestnut. (Is it me, or does it appear lately that fewer children are practising the age-old game of 'conkers'? I was very proud to have a sixer at one stage in my conkering career.) Conkers form inside spiny green capsules after the candle-like blooms have been pollinated by bees.

Because of its size this really is not a tree for even an average-sized garden but I have included it in the gazetteer because it is such a beneficial plant to bees.

Crataegus monogyna (Hawthorn, quickthorn or may)
(*Rosaceae*)

Good for	Nectar and pollen
Hardy	
Deciduous	
Height	Up to 6m
Flowers	White; late April, May
Best position	Full sun, partial shade
Soil	Any except waterlogged

Hawthorn must be one of the least fussy and most versatile trees that we have. It will grow in just about any location, including coastal and exposed areas, and, as well as making a beautifully shaped tree, it can be grown as a hedge. It is a joy to behold in spring when it is covered with white flowers, which, when pollinated, turn pink and produce abundant red berries in the autumn. Some cultivated varieties have pink or even red flowers, but beware of those varieties with double flowers; these are of no use to bees.

There is an old saying, ''Ne'er cast a clout till May be out', which means that you should not go without your outdoor clothing until May is over. Some people regard this as being the month of May, but others think it refers to the flower of the may which is in blossom in mid– to late spring.

May blossom was also used to garland the May Queen (I cannot help but think it would be a bit too prickly) and it has been surmised that the old rhyme 'Here we go gathering nuts in May' is a corruption of 'Here we go gathering knots of may', meaning collecting bunches, or 'knots', of may blossom for the celebrations. Some people, however, deem it bad luck to actually bring may blossom into the house.

Whatever the tales surrounding it, hawthorn provides pollen and varying amounts of nectar for our bees and I, for one, look on it as the harbinger of summer.

Spring/summer

Ilex aquifolium (Holly)

(*Aquifoliaceae*)	
Good for	Nectar and pollen
Hardy	
Evergreen	
Height	Up to 20m
Flowers	Insignificant white flowers; May and June
Best position	Will grow in sun or shade
Soil	Moderately fertile, moist but well drained

Ilex aquifolium is the common holly, which is technically a tree but is usually found as a shrub. They do not mind being pruned so are ideal specimens to be shaped into topiary. There are dozens of different varieties, many of them variegated, all of which are either male or female, so if you want berries on your female plant you must be sure to have a male plant nearby for the bees to visit too. A favourite of mine is *Ilex aquifolium* 'Silver Queen' (which is a little confusing as it is a male plant), which has lovely dark green foliage with a cream edge. Very similar is *I. aquifolium* 'Argentea Marginata' (which is female) so I have both in my garden to guarantee some berries.

Sorbus aucuparia (Rowan or mountain ash)

(*Rosaceae*)	
Good for	Nectar and pollen
Hardy	
Deciduous	
Height	Ultimately 12m
Flowers	White; May to June
Best position	Full sun or partial shade
Soil	Almost any type except waterlogged

The rowan is a very versatile tree, growing happily in cities as well as in the countryside and at high altitudes. It is reputed to have magical powers and can often be seen in churchyards to stop ghosts disturbing the living, and near homes and on farms to protect people and livestock from witches. On a more practical note, the berries are high in Vitamin C and, although bitter and all but inedible raw, when used in jam or jelly they make a traditional accompaniment to game dishes. Birds love the berries too and, of course, bees love the flowers.

────────────── *Summer* ──────────────

Castanea sativa (Sweet chestnut or Spanish chestnut)
(*Fagaceae*)

Good for	Nectar and pollen
Hardy	
Deciduous	
Height	Up to 15m
Flowers	Catkins; late June/ July
Best position	Full sun
Soil	Free draining

This is not to be confused with horse chestnut, which is an entirely different species. This chestnut, which is thought to have been introduced into Britain by the Romans, is best known for its edible fruits, which can be roasted, used in confectionery, ground to make flour, and are the main ingredient of stuffing for the Christmas turkey!

The catkins contain the flowers of both sexes with the female flowers in the lower part and the male in the upper. Once pollinated the female flowers produce brown nuts which are protected by spiny outer cases, designed to deter squirrels.

Catalpa bignonioides (Indian bean tree) (page 87)
(*Bignoniaceae*)

Good for	Nectar and pollen
Hardy	
Deciduous	
Height	Up to 15m
Flowers	White with yellow and purple flecks; July/August
Best position	Full sun, sheltered
Soil	Well drained/light, moist

Although not native to Britain, having been introduced from the United

States in 1726, it has become a favourite in large gardens here, not least because of its large clusters of flowers. Once pollinated (by bees, naturally!), thin, green, bean-like pods (which turn brown) are formed, giving rise to part of its common name.

There is a much smaller variety, 'Nana', which reaches about 3.5 metres – much better for smaller gardens.

Tilia (Lime or linden)
(*Tiliaceae*)

Good for	Nectar and pollen
Hardy	
Deciduous	
Height	Up to 35m, depending on type
Flowers	Pale yellow; July/August
Best position	Full sun
Soil	Any except waterlogged

First of all be sure *not* to grow *Tilia petiolaris, T. oliveri* or *T. orbicularis* – these are hazardous to bees.

Lime trees have had some bad press in recent years. The once ubiquitous street tree, the common lime (*Tilia vulgaris*), has fallen out of favour for several reasons: first, because of the honeydew secreted from aphids which feed on it and the subsequent mess; second, because it loses its leaves relatively early; third, because it suckers prodigiously, which means extra maintenance (and therefore cost); and fourth, its size. No wonder it is doomed.

There are other limes, however, which can be grown and which bees will flock to, the most obvious ones being the small-leaved lime (*Tilia cordata*) and the broad-leaved lime (*T. platyphyllos*). There is another, the Mongolian lime (*T. mongolica*), which is well worth growing: it is much smaller than many other limes, reaching a height of 15m, but only after 20 years or so.

The individual flowers of lime open at night and last for about a week; nectar is secreted mainly during the morning throughout the flowering time. In many European countries the flowers are collected and dried and used for making tea or a tisane.

Tetradium daniellii (Chinese bee tree or bee-bee tree)
(*Rutaceae*)

Good for	Nectar and pollen
Hardy	
Deciduous	
Height	Up to 15m
Flowers	Creamy white; August
Best position	Full sun or light, dappled shade
Soil	Well drained but moist

As you can guess from its common names, this tree is very attractive to bees! Its broad heads of flowers appear late in the season, providing a welcome source of food for our buzzy friends. Part of its Latin name commemorates a certain William Daniell who, in the 1860s, collected specimens of the tree in the Tientsin province of China.

—————————— *Summer/autumn* ——————————

***Sophora japonica*,** syn. **_Styphnolobium japonicum_** (Japanese pagoda tree or Chinese scholars' tree)
(*Fabaceae*)

Good for	Nectar and pollen
Hardy	
Deciduous	
Height	Up to 15m
Flowers	Panicles of creamy white flowers; August/September
Best position	Full sun
Soil	Light, well drained

Despite its name implying that the pagoda tree comes from Japan it is in fact native to China. It found its way to Japan and then, in 1753, to Britain. It makes a beautiful, fairly large tree, with a profusion of creamy white flowers held on panicles, which are extremely useful for bees as they appear late in the season.

Flowering times of selected trees

	Feb	Mar	Apr	May	Jun	Jul	Aug	Sep
Corylus avellana								
Salix caprea								
Aesculus hippocastanum								
Crataegus monogyna								
Ilex aquifolium								
Sorbus aucuparia								
Castanea sativa								
Catalpa bignonioides								
Tilia sp.								
Tetradium daniellii								
Sophora japonica								

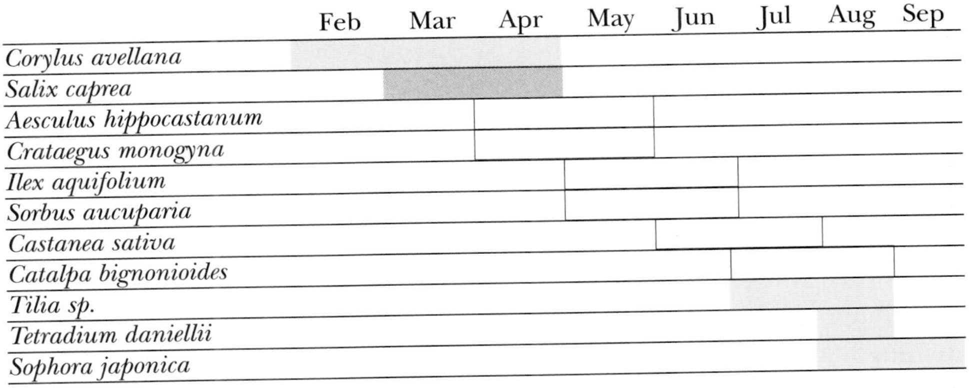

Figure 14 Flowering times of selected trees

Edible fruits

I have decided to create a separate list of edible fruits. They seemed a little out of place tucked in among the trees and shrubs and they are so dependent on the good auspices of our buzzy friends that I think they deserve a section of their own. This is by no means an exhaustive list; it is a selection of a few fruiting plants that you may want to grow in your own garden.

Spring

Ribes uva-crispa var. reclinatum (Gooseberry)

(*Rosaceae*)	
Good for	Nectar and pollen
Hardy	
Deciduous	
Height	1.5m
Flowers	Small, white; March/April
Best position	Sun or partial shade, sheltered
Soil	Does best in fertile, well-drained but moisture-retentive soil

Gooseberries are one of the first bush fruits to flower in the spring so some fruits are ready to be picked just when the elderflower comes into blossom. The blend of gooseberry and elderflower is really a marriage made in heaven, although I'm not sure how or when the two got together. Suffice it to say, if you have not tried this sublime combination, get hold of (or better still, make) some jam, fool or curd and I can guarantee you will not want to share it.

There are several varieties worth growing but perhaps the best-known is 'Careless', which was introduced in 1855 and is still popular. It is a 'traditional' variety bearing green fruit, but not all gooseberries are green: 'Whinham's Industry' is a red-fruited variety, as is 'Lancashire Lad' (I had to include it!) which dates from 1824.

If there is a geographical area traditionally associated with gooseberries then it has to be Lancashire and Cheshire (just as rhubarb is associated with Yorkshire). During the 1860s gooseberry clubs sprang up in these areas and there was fierce competition for who could grow the finest, biggest berries.

Pruning gooseberries to keep them fruiting is fairly straightforward: during the dormant winter season cut out any dead or diseased stems and any crossing over the centre. The idea of the latter is to keep the centre of the bush open to light and air to minimize the risk of disease. If some of the branches are becoming too long, shorten them by about half, cutting back to a young shoot.

Prunus domestica (Plum including greengages)

(*Rosaceae*)

Good for	Nectar and pollen
Hardy	
Deciduous	
Height	2 – 4.5m, depending on rootstock
Flowers	Pink; March/April
Best position	Full sun
Soil	Well drained but moisture retentive

Usually when you ask someone to name some varieties of plum they say 'Victoria' and then grind to a halt. On one nursery website I found 75 different plums with wonderful names such as Warwickshire Drooper, Sanctus Hubertus and Guthrie's Late Green. Some varieties start flowering as early as late March, while others still have not flowered some two to three weeks later. The blossom appears before the leaves and will last about a week. Although many plum varieties are self-fertile (which means you do not need another variety nearby to ensure pollination), the yield will improve considerably if bees work the flowers.

As with other fruiting trees you can decide how big you want your tree to grow by choosing an appropriate rootstock to which the fruiting top part has been grafted – for plums these range from semi-dwarfing to very vigorous, so any size of garden can accommodate a plum tree.

Malus domestica (Apple)

(*Rosaceae*)

Good for	Nectar and pollen
Hardy	
Deciduous	
Height	1.2 – 4.5m, depending on the rootstock
Flowers	White/pale pink; April/May
Best position	Full sun
Soil	Moist, but well drained

Everyone can grow an apple tree in their garden, no matter how small or large it is. (Actually, I should say 'two apple trees', since they need to be cross-pollinated by another cultivar flowering at the same time to achieve a crop of fruit.) Even if your garden is very small this should not pose too much of a problem because you can even grow apple trees in containers. This is possible because the eventual size of the tree will depend on the rootstock to which the fruiting top part has been grafted. For example an M27 rootstock will produce a tree with a mature height of about 1.2m, whereas with an M25 rootstock you will get a tree up to 4.5m high.

Your tree does not necessarily have to be 'tree-shaped' either: you can train it into different shapes so that it can be grown against a wall, for example, or even by the side of a path so that you can step over it.

There are hundreds of varieties to choose from too. You can have eating or cooking apples, or ones that double up as both. Your local climate will dictate to a certain degree which varieties you can grow successfully so it is worth contacting a specialist nursery to ask for their advice.

Whatever tree you choose it will have to be pruned in order to get the best crop but that is a subject that is beyond the scope of this book – there are numerous texts which will give sound advice.

Pyrus communis (Pear)
(*Rosaceae*)

Good for	Nectar and pollen
Hardy	
Deciduous	
Height	2 – 6m, depending on the rootstock
Flowers	White; April/May
Best position	Full sun
Soil	Moist, but well drained

Much of what I have said about apples also applies to pears: you need to decide how big a tree you want, what shape you want it to be, and if you want a pear for eating or cooking. I have two lovely varieties in my garden, a 'Black Worcester', which is a historic culinary variety dating back to 1575, and a dessert one called 'Durondeau', which originates from Belgium. You can even get varieties from which to make perry (a little like cider but made from pears). As with all other fruiting trees, pears need bees for pollination, so that in itself is a good enough reason to grow them.

Prunus avium (Sweet cherry)
(*Rosaceae*)

Good for	Nectar and pollen
Hardy	
Deciduous	
Height	3 – 6m, depending on rootstock
Flowers	White; April/May
Best position	Full sun
Soil	Well drained but moisture retentive

Like other fruit trees there are numerous varieties of edible cherry which have been cultivated for increased size and sweetness and less disease. A couple of British-raised sweet varieties are 'Penny' and 'Merton Premier', which are just as cherries should taste; but beware – birds like them too so you will have to keep a good lookout and gather them as soon as they are ripe enough to eat. The most common acid cherry, which is suitable only for culinary use, is 'Morello', but there are others available, such as 'Kentish Red' and 'Montmorency'. Be careful when you buy your tree to make sure which type it is otherwise you could end up with rather more tart-tasting fruit than you expected.

Ribes nigrum, R. rubrum (Blackcurrant, red and white currant)
(*Grossulariaceae*)

Good for	Nectar and pollen
Hardy	
Deciduous	
Height	1.2m
Flowers	Small, white, yellow or green flowers on drooping stems; April/May
Best position	Sun or partial shade, sheltered
Soil	Does best in fertile, well-drained but moisture-retentive soil

Currants are members of the same family as gooseberries (in fact the gooseberry has been crossed with a blackcurrant to produce the jostaberry (*Ribes* x *culverwellii*)) and so require the same sort of growing conditions.

There are three types of common currants, black, red and white, although the white currant is actually a variety of the redcurrant, *Ribes rubrum*, and not a separate species at all. All currants are ideal for use in pies, jams and puddings, especially the scrumptious dessert Summer Pudding, which makes the most of soft fruits and berries and is even better when served with a dollop of clotted cream!

One of the best blackcurrants is 'Ben Connan', an early-fruiting variety, which has really large berries and a lovely, rich flavour. A late cropper is Ben Sarek. Of the redcurrants, 'Jonkheer van Tets' is a good early variety, with 'Rovada' for a late crop. And of the few white currants available, 'White Grape' and 'White Versailles' are the best.

Although currants will bear fruit on older wood, the amount and quality will decline over the years. To ensure that new growth is available year-on-year, you should plant the bushes deeply so that new stems appear each year from below soil level. This way you can remove about a third of the oldest stems each year to ground level, and still have sufficient fruiting branches for the following season.

———————————— *Spring/summer* ————————————

Mespilus germanica (Medlar) (page 197)

(*Rosaceae*)

Good for	Nectar and pollen
Hardy	
Deciduous	
Height	3 – 6m, depending on rootstock
Flowers	White, often touched with pink; May to June
Best position	Sun or partial shade, sheltered
Soil	Does best in fertile, well-drained but moisture-retentive soil, but can tolerate poorer soil

It is thought that the medlar originated in the eastern Mediterranean. What we do know is that the Romans cultivated them, and probably brought them to Britain.

The fruit is ready to pick in late October or early November when the stalk parts easily from the tree. At this stage they will still be hard so they must be allowed to soften before they can be eaten. This process is called 'bletting', and takes from two weeks to a month. The fruit is ready when the hard, cream-coloured flesh turns brown and mushy, but not rotten – it has been said that they are 'ripened by their own corruption'. At this stage they can be eaten raw, but they can also be used in jellies, desserts and preserves.

A rather more descriptive name for medlar is 'dog's bottom', a translation of the French *cul de chien*. Looking at the fruit from a particular angle, you can see why! (Chaucer, in *The Reeve's Tale*, refers to them in an even more colourful way.)

Mespilus germanica (medlar) (opposite)

Cydonia oblonga (Quince)

(*Rosaceae*)

Good for	Nectar and pollen
Hardy	
Deciduous	
Height	2.5–6m, depending on rootstock
Flowers	Pink or white; May or early June
Best position	Sun, sheltered
Soil	Fertile, well drained but moisture retentive

The origin of the quince is a little uncertain but it is thought to have made its way to Britain from southern Europe. Indeed, the first records of its cultivation here date from 1275. It was very popular from the 16th to 18th centuries, especially as a herbal medicine, treating a whole range of maladies from sore throats and diarrhoea to inflammation. In 1653 Nicholas Culpeper the herbalist also advocated its use for those who were a little hirsutely challenged: made up into a plaster with wax 'it brings hair to them that are bald, and keeps it from falling off'. Mmm – I am a little sceptical about that one.

Be that as it may, quince was also widely used in the kitchen, its fragrant flesh being added to fruit pies, preserved as jam or simply stewed. It fell out of favour for some time, however, but recently it seems to be making somewhat of a comeback and I have even seen fresh pear-shaped fruits on sale in a local supermarket. Quince can be eaten raw but it is hard and gritty and is much better left to soften or, better still, cooked.

Throughout continental Europe you will find variations of quince paste or jam – *Membrilo* in Spain, *Cotignac* in France, *Marmelada* in Portugal (from which our word for orange 'jam' is derived).

There are 11 varieties of quince listed in one nursery catalogue – but do not confuse them with *Chaenomeles japonica*, also known as quince, but grown for purely ornamental purposes.

Rubus idaeus (Raspberry)

(*Rosaceae*)

Good for	Nectar and pollen
Hardy	
Height	Up to 1.5m

Flowers	White; May to July, depending on variety
Best position	Sun, sheltered
Soil	Fertile, moisture retentive, slightly acidic

The raspberry is a perennial plant which generally produces biennial canes. This means that in its first year the cane simply grows; in its second year it produces flowers (pollinated by bees, of course!) and subsequent fruit. At any one time during the growing season the plant will have new, non-fruiting canes and second-year fruiting canes on the same plant. (There are exceptions to the rule, such as the variety 'Autumn Bliss', which carries fruit on the current season's canes.) After the fruit has been harvested, the cane should be cut out at the base.

There are numerous varieties of raspberry available. Some flower early and therefore produce an early crop (for example, 'Glen Moy'), others flower and fruit later (such as 'Glen Prosen', followed by 'Glen Magna'). Some, like 'Allgold', have yellow fruit.

Raspberries need a lot of food to produce good fruit, so they should be given a slow-release, general fertilizer in spring followed by a good mulch of well-rotted organic matter. They should also be watered well in dry periods.

Rubus fruticosus (Blackberry) (page 25)

(*Rosaceae*)	
Good for	Nectar and pollen
Hardy	
Height	Up to 1.8m
Flowers	White or pale pink; May/June, but can have an additional flush later
Best position	Sun, sheltered
Soil	Fertile, on the light side

When I hear the name blackberry I always think of blackberry and apple pie made from the bramble berries that my sister and I used to pick in the hedgerows in autumn – and of all the scratches that we ended up with while picking them! I still pick from the hedgerow, for tradition's sake more than anything else because I have a cultivated variety of blackberry in the garden which yields much larger fruits, and is thornless to boot! The one I grow is 'Oregon Thornless' which, as well as producing a good crop of berries, is also quite decorative with its large, white blossom which the bees love, and its parsley-shaped leaves. It is also ideal for a smaller garden because it is fairly compact – some would say weak – but that's fine.

Pruning is the same as for raspberries, so the general rule of thumb is that when the cane has fruited it should be cut out at ground level. Any new shoots that have not yet borne fruit should be tied to supporting wires to stop them flopping over.

Fragaria x *ananassa* (Strawberry) (page 21)
(*Rosaceae*)

Good for	Nectar and pollen
Hardy	
Height	45cm
Flowers	White; May to August
Best position	Sun, sheltered
Soil	Fertile, on the light side

Strawberries are perhaps one of the best-loved of all soft fruits. Their taste is redolent of summer – and where would Wimbledon be without strawberries and cream?

There are many varieties to choose from: some produce fruits as early as June, others as late as September, but probably the best-known is 'Cambridge Favourite'. It crops reliably under a wide range of growing conditions so it is not surprising that it has been given the Royal Horticultural Society's Award of Garden Merit. Another variety that has been awarded this honour is 'Alice', which crops later than 'Cambridge Favourite'.

Although you do not actually prune strawberries, they do produce runners during the growing season which bear young plantlets. If you do not want to propagate new plants then these should be cut off to conserve the original plant's energy. If, however, you do want to renew your stock then pop the young plantlets into pots, still attached to the 'mother' plant, until they have formed roots of their own. You can then cut the runner and grow them on.

I like to grow some alpine strawberries, too. These are a different species from the 'garden' ones and go under the name of *Fragaria vesca*. The fruits are much smaller but, to my mind, tastier, and they keep flowering and cropping throughout the season – lovely for me and the bees! If I have enough fruits I like to make some jam, replacing some of the sugar with honey – it really is the finest strawberry jam I have ever tasted.

Flowering times of selected edible fruits

	Mar	Apr	May	Jun	Jul	Aug
Ribes uva-cripa var. reclinatum (Gooseberry)						
Prunus domestica (Plum)						
Malus domestica (Apple)						
Pyrus communis (Pear)						
Prunus avium (Cherry)						
Ribes nigrum, R. rubrum (Currants)						
Mespilus germanica (Medlar)						
Cydonia oblonga (Quince)						
Rubus fruticosus (Blackberry)						
Rubus idaeus (Raspberry)						
Fragaria x ananassa (Strawberry)						

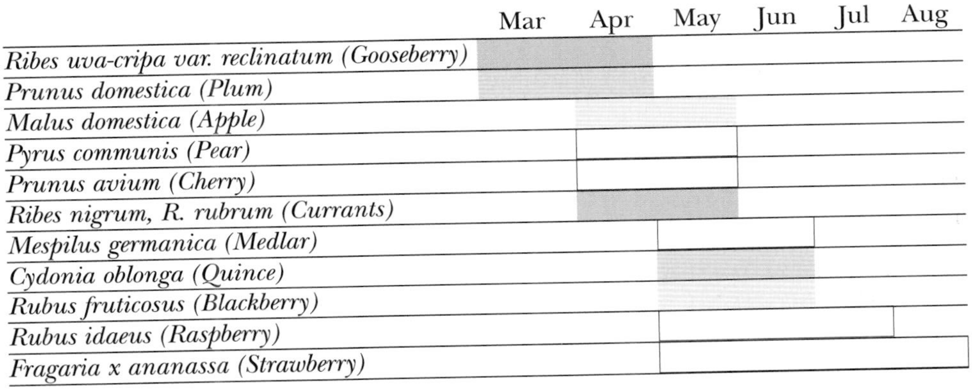

Figure 15 Flowering times of selected edible fruits

Herbs

Spring

Laurus nobilis (Bay)

(*Lauraceae*)	
Good for	Nectar
Type of plant	Perennial evergreen tree
Height	Up to 8m
Flowers	Pale yellow; March to May
Best position	Warm, sunny site. Protect from cold, dry winter winds
Soil	Fertile, well drained

Bay is said to have originally come from Asia Minor but it grows extensively in Mediterranean regions. The Greeks and Romans saw bay as a symbol of wisdom and glory, and the laurel wreaths that were given to victorious statesmen and athletes were in fact wreaths of bay. It is thought to have been introduced into Britain in 1562.

Bay is best grown in fertile, well-drained soil in a sheltered, sunny position. It may need protection in winter in the form of a layer or two of horticultural fleece. It also makes an excellent container plant which can be moved into an unheated greenhouse or conservatory for the winter months.

Bay is an excellent subject for topiary (which is why I have included a 'lollypop' one in the Herb Garden Planting Plan) but if left to its own devices it can eventually reach 8m high with a spread of 3m.

The flowers may be insignificant and easily overlooked but they produce a lot of nectar which during the early part of the season is a boon to bees.

As with many herbs the leaves of the bay can be used in the kitchen and form an essential ingredient in a bouquet garni. But beware: *Laurus nobilis* is the only member of the laurel family whose leaves are edible.

Rosmarinus officinalis (Rosemary)
(*Lamiaceae*)

Good for	Nectar and pollen
Type of plant	Hardy evergreen shrub
Height	Up to 1m
Flowers	Pale blue; March to May
Best position	Sunny with shelter from cold winds
Soil	Well drained

Rosemary is native to scrubby, coastal regions of the Mediterranean – its Latin name, *Rosmarinus*, means 'dew of the sea'. It also grows abundantly inland: during the 16th century, the French gardener Olivier de Serres noted that in Provence rosemary was so abundant that its woody stems were used as fuel in bread ovens. Rosemary is a symbol of fidelity and remembrance, and is beloved by bees – Sir Thomas More (1478–1535) wrote: 'As for Rosemarine [*sic*], I let it run all over my garden walls, not only because my bees love it, but because 'tis the herb sacred to remembrance, and, therefore, to friendship.'

Being a Mediterranean herb it enjoys a warm, sunny position. Although it is hardy and can withstand temperatures down to −10°C and lower in a sheltered position, the one thing it really hates is wet feet. Give it well-drained, verging on poor, soil.

The famous Narbonne honey from the Aude *département* of south-western France is rosemary honey, traditionally harvested from the hives on Midsummer's Day, 24 June. It is a very light-coloured honey, and infused with the scent of rosemary. It is also very expensive, even in France.

———————————— *Spring/summer* ————————————

Allium schoenoprasum (Chives)

(*Alliaceae*)

Good for	Nectar and pollen
Type of plant	Hardy perennial
Height	30cm
Flowers	Purple/pink; May to September
Best position	Preferably sunny but dappled shade is tolerated
Soil	Rich, moist

This herb is a member of the onion family and can be found growing wild in Europe, Asia and North America. Although they have been gathered from the wild since antiquity it seems probable that they were not cultivated until the Middle Ages.

Chives like rich, moist soil which must not be allowed to dry out during the growing season. They love a warm, sunny position, but will tolerate dappled shade. They can also be grown in a container, but should be divided in the autumn if the pot gets overcrowded.

The pink 'drumstick' flowers, beloved by bees, are edible, and look – and taste – good divided into florets in a summer salad.

———————————— *Summer* ————————————

Foeniculum vulgare (Fennel)

(*Apiaceae*)

Good for	Nectar and pollen
Type of plant	Hardy perennial (but often grown as an annual)
Height	Up to 2m
Flowers	Yellow; June to August
Best position	Sun
Soil	Fertile, moist but well drained

From a cook's point of view the main reason for growing fennel is for its leaves. This is true of *Foeniculum vulgare* but there is another fennel, *Foeniculum vulgare* var. *dulce* (Florence fennel) which is primarily grown for its bulbous root.

From a bee's point of view it is the flowers that are of interest, although you will find that it is often overrun with more short-tongued insects seeking out its liquorice-flavoured nectar and bees will look elsewhere. If you do let it flower be sure not to grow it near *Anethum graveolens* (dill) as they will hybridize and their offspring are of no value either as herbs or ornamentals.

Folklore has it that if you hang a bunch of fennel above your door on Midsummer's Eve then you will be protected from enchantment and witches.

Fennel likes moist, fertile soil and a sunny position. It looks magnificent if grown in a herbaceous border, where its feathery leaves make a superb foil for more robust specimens.

Lavandula (Lavender) (page 88)
(*Lamiaceae*)

Good for	Nectar and pollen
Type of plant	Hardy evergreen shrub
Height	Up to 90cm, depending on type
Flowers	Purple; June to August
Best position	Sunny
Soil	Well drained

Lavender is native to dry, rocky regions of the Mediterranean. Its name comes from the Latin *lavare*, meaning 'to wash', and Romans used lavender to scent their bath water. It is thought that it was indeed the Romans who brought lavender to our shores, although the first reference to it appears in 1265 when it is recorded in a manuscript of that date. Although there are many species of lavender the hardiest are *Lavandula angustifolia* and *Lavandula* x *intermedia*, which can withstand temperatures down to −10°C and lower provided they are not exposed to cold winds and they do not have wet feet.

Bees love lavender. Nectar is produced freely and is readily accessible by all bees since the corolla tube of the flower is, on average, 6mm deep – well within the reach of even honeybees. Where bees feed exclusively on lavender the honey will be quite dark and when it granulates it has a very fine grain, almost as smooth as butter. Lavender used to be grown in

England on quite a large scale to supply perfume houses with oil, but the number of lavender fields declined as synthetic essences took over. Over recent years, however, there has been a renewed interest in natural oils and lavender fields can now be found in areas such as Cheshire, the Cotswolds, Kent, Norfolk and Yorkshire. Perhaps it will not be long before there is English lavender honey for sale too.

Melissa officinalis (Lemon balm)

(*Lamiaceae*)	
Good for	Nectar
Type of plant	Hardy perennial
Height	30cm
Flowers	Cream; June to August
Best position	Preferably sunny but dappled shade is tolerated
Soil	Moisture retentive – but not waterlogged

Lemon balm is native to southern Europe and has been cultivated for over two thousand years: it was brought to Britain by the Romans. Melissa is the Greek word for honeybee and the herb has long been associated with bee-keeping. Not only is it a rich source of nectar, but Virgil (70–19 BC) notes that balm induces swarming, and in the 16th century the herbalist John Gerard wrote that: 'The hives of bees being rubbed with the leaves of balm causeth the bees to keep together and causeth others to come with them. When they are strayed away, they do find their way home by it.'

Lemon balm is not as fussy as some herbs and will grow in both sun and dappled shade. It prefers moisture-retentive soil but will not tolerate permanently wet feet.

Origanum vulgare (Wild marjoram or oregano) (page 25)

(*Lamiaceae*)	
Good for	Nectar
Type of plant	Hardy perennial
Height	Up to 45cm
Flowers	Pale mauve; June to August
Best position	Full sun
Soil	Well drained

There are several species of *Origanum*, which often causes confusion since the common name is sometimes oregano and sometimes marjoram. The 'wild' one is *Origanum vulgare* but you may also find *Origanum marjorana* (sweet marjoram), which is less hardy and has white flowers, and *Origanum onites* (pot marjoram), which has pink/purple flowers. All species of *Origanum* appear to originate from the Mediterranean but nowadays many species can be found growing wild in various places, including the chalk downs of Britain. Like most herbs, they all need a sunny site with well-drained soil.

Origanum vulgare is one of the best herbs to grow for bees. The sugar content of the nectar is one of the highest among all plants and the resulting honey is of top quality.

Mentha spicata (Garden mint) (page 68)
(*Lamiaceae*)

Good for	Nectar
Type of plant	Hardy perennial
Height	Up to 60cm
Flowers	Purple/mauve; June to August
Best position	Sunny or light shade
Soil	Fertile, moist

There are numerous species and varieties of mint but the most common garden mint to be found in Britain is spearmint (*Mentha spicata*), although the named variety *Mentha spicata* 'Tashkent' has the best flavour. It is native to Mediterranean regions and was probably introduced to Britain by the Romans. In classical times mint was rubbed over tables before a banquet so that, as Pliny noted, the scent would 'stir up … a greedy desire for meat'.

Another mint which you will find readily available in garden centres, that is every bit as good as *Mentha spicata*, is *Mentha suaveolens* (apple mint). This too is hardy but it has larger, hairier, softer green leaves and can grow up to 20cm taller.

All types of mint are invasive, so grow it either in a container or in a bed of its own. Although it prefers a sunny site it will grow in light shade as long as it has fairly fertile, moist soil.

Thymus vulgaris (Common or garden thyme)

(*Lamiaceae*)	
Good for	Nectar
Type of plant	Hardy, evergreen perennial
Height	Up to 45cm
Flowers	Pale pink; June to August
Best position	Warm and sunny, sheltered
Soil	Well drained

According to the Royal Horticultural Society there are about 350 species of thyme, including some which grow wild in Britain. Three that are widely grown are garden or common thyme (*Thymus vulgaris*), lemon thyme (*Thymus* x *citriodorus*) and broad-leaved thyme (*Thymus pulegioides)*. They are native to rocky sites and dry, usually chalky, grasslands of Europe, western Asia and North Africa.

Thyme likes a warm, sunny site with well-drained soil and it hates wet feet, particularly during the winter.

Lemon thyme adds a subtle hint of citrus in many dishes and I have even found a recipe for honey and lemon thyme ice cream.

Satureja montana, *Satureja hortensis* (Winter savory, summer savory)

(*Lamiaceae*)	
Good for	Nectar
Type of plant	Hardy, semi-evergreen perennial; annual
Height	30cm
Flowers	White with pink-mauve tinge; June to August
Best position	Full sun
Soil	Poor, well drained

There are basically two types of savory – *Satureja montana*, winter savory, and *Satureja hortensis*, summer savory. Both bear small white flowers during the summer and both are attractive to bees. The difference between them is that the winter savory is a perennial and can be kept over winter, whereas summer savory is an annual and must be resown each year. For this reason, I tend to grow only the winter savory.

Savory comes from the Mediterranean regions where it grows in stony or sandy fields or along the roadside, which gives a foolproof pointer to the sort of conditions it likes.

Savory is best grown in full sun and winter savory, being a semi-evergreen perennial, will hold a good deal of foliage over the winter.

────────────── *Summer/autumn* ──────────────

Hyssopus officinalis (Hyssop) (page 67)
(*Lamiaceae*)

Good for	Nectar and pollen
Type of plant	Hardy, semi-evergreen perennial
Height	80cm
Flowers	Blue, pink or white, depending on variety; June to September
Best position	Sunny
Soil	Well drained

Hyssop is native to southern Europe, the Near East and southern Russia. Its use goes back to antiquity when it was used more as a medicinal or purifying herb than a culinary one. Hyssop could always be found in monastery gardens where it was grown as a 'cure-all' for diverse complaints from acne to worms. Indeed, Gerard, in his herbal of 1597, left hyssop 'altogether without description, as being a plant so well knowne that it needeth none'. The same is not true today since it figures only rarely in modern recipes.

Hyssop is a sun worshipper and must have well-drained soil. It makes an ideal low hedge: trim back the flowers (you can get blue, pink or white flowered varieties) in autumn to keep them in shape. It also makes a perfect container plant.

Salvia officinalis (Sage)
(*Lamiaceae*)

Good for	Nectar
Type of plant	Hardy, evergreen perennial
Height	Up to 60cm
Flowers	Mauve-blue; June to September
Best position	Warm and sunny
Soil	Well drained – not acid

The genus *Salvia* contains a vast number of species (more than 700) including the tender *Salvia splendens* often used as summer bedding. We are concerned with the herb sage, which has always been highly prized, not only in the kitchen but also as a medicine – its name is testament to this: the Genus name 'Salvia' is said to be derived from the Latin *salvere*, meaning to save or heal, and the species name 'officinalis' comes from the Latin word *opificina*, meaning a herb store or pharmacy.

An Arabic proverb asks: 'How can a man die who has sage in his garden?' And the Chinese valued sage so highly as a medicinal herb that in the heyday of the tea trade with the Dutch, when tea commanded a premium price, one pound (weight) of dried sage leaves would be traded for three pounds of tea.

Salvia officinalis (known as common, garden or broad-leaved sage) is the best-known sage for culinary use, with pale grey-green, velvety leaves. However, there are other varieties, notably Gold Sage (*Salvia officinalis* 'Icterina') or Purple Sage (*Salvia officinalis* 'Purpurascens' group) with colourful leaves, which are just as good for bees as the common sage. Sage requires a sunny position with well-drained soil. Although it is a Mediterranean plant, it will survive British winters without protection, as long as its feet are not wet.

Flowering times of selected herbs

	Mar	Apr	May	Jun	Jul	Aug	Sept
Laurus nobilis							
Rosmarinus officinalis							
Allium schoenoprasum							
Foeniculum vulgare							
Lavandula sp.							
Melissa officinalis							
Origanum vulgare							
Satureja montana							
Mentha spicata							
Thymus vulgaris							
Hyssopus officinalis							
Salvia officinalis							

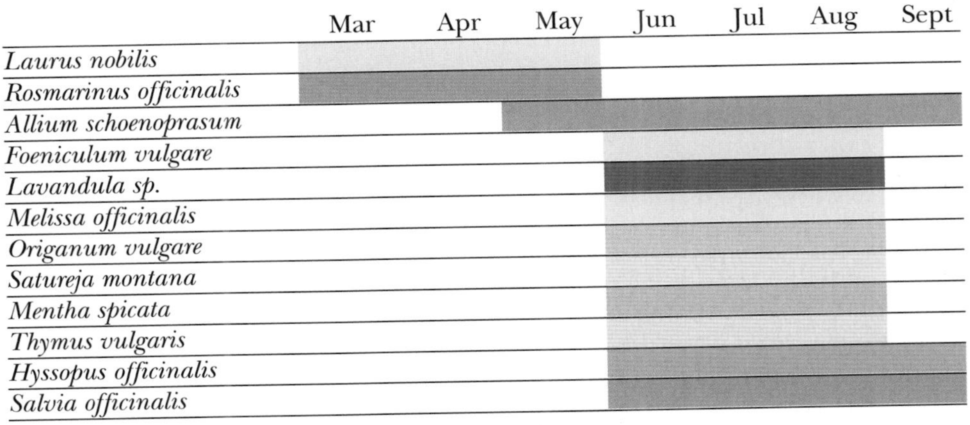

Figure 16 Flowering times of selected herbs

Wild flowers

Spring/summer

Tussilago farfara (Coltsfoot)

(*Asteraceae*)	
Good for	Nectar and pollen
Type of plant	Perennial
Height	Up to 10cm
Flowers	Yellow; February to April

Coltsfoot is one of the first wild species to flower in the spring, appearing at about the same time as crocuses. Some shoots develop flowers, the seeds of which are topped with a tuft of silky 'hairs' which are often used by goldfinches to line their nests. Other shoots develop leaves which appear after the flower stems have died down in May. Coltsfoot can be invasive, but it is such a good early bee plant that it is worth putting up with its rampant nature.

Lamium purpureum/L. album (Red/white dead nettle) (page 211)

(*Lamiaceae*)	
Good for	Nectar and pollen
Type of plant	Annual
Height	Up to 45cm
Flowers	Dark pink-purple/white; February/March to October, sometimes later

This is not a true nettle but its leaves are enough like one to make people wary. Although there are many attractive cultivated varieties which deserve a place in the garden, the 'original' version is thought of as a pernicious weed.

Taraxacum officinale (Dandelion)

(*Asteraceae*)	
Good for	Nectar and pollen
Type of plant	Perennial
Height	5–30cm
Flowers	Yellow; March to October

Lamium album (white deadnettle)

This yellow, daisy-like flower takes its common name, dandelion, from the French *dents de lion*, meaning lion's teeth, which refers to the shape of the leaf. Despite being reviled as a weed by gardeners it is one of the best 'bee plants', giving up both pollen and nectar. It can be kept in check by constant deadheading.

Ornithogalum pyrenaicum (Bath asparagus)

(*Liliaceae*)

Good for	Nectar and pollen
Type of plant	Bulb
Height	Up to 80cm
Flowers	Greenish-white flowers on spikes; April to June

The common name of this plant gives us a clue to where it is naturally found, and also to its use. It has a limited distribution in Britain, being found mostly in the Avon valley around Bath, although it is widespread throughout Europe. One reason that has been put forward for its localized occurrence is that the Romans brought it with them as a food crop when they occupied areas in and around Bath. It can be eaten like asparagus before the flower spikes open. Although it is scarce in the wild, bulbs can be bought from reputable suppliers.

Leucanthemum vulgare (Ox-eye daisy, moon daisy) (page 37)

(*Asteraceae*)

Good for	Nectar and pollen
Type of plant	Perennial
Height	Up to 60cm
Flowers	White, daisy flowers; May to September

If you view this lovely plant in the half-light of a summer's evening you can understand why one of its common names is moon daisy: the blooms really shine out like miniature moons. It can often be found on edges of fields and verges, especially where ground has been newly sown with conservation flora after civil engineering works and the like.

The garden perennial daisy, or Shasta daisy, *Leucanthemum* x *superbum*, is a related hybrid.

Borago officinalis (Borage or starflower) (page 24)

(*Boraginaceae*)	
Good for	Nectar and pollen
Type of plant	Annual
Height	30–60cm
Flowers	Blue; May to September

Originally a Mediterranean plant, borage can now be found naturalized across northern Europe. It is often found in the herb garden (the young leaves and flowers can be eaten and are traditionally part of a refreshing Pimm's) and it is grown commercially for its oil. Its habit is a little unruly, however, and so it is ideal for a more relaxed or 'wild' planting scheme. Bees adore it.

Trifolium repens (White clover) (page 25)

(*Fabaceae*)	
Good for	Nectar and pollen
Type of plant	Perennial
Height	10cm
Flowers	White (sometimes with a tinge of pink); May to October

The Latin name tells us a lot about this plant: *trifolium* means it has three leaves, and *repens* means creeping. Like many other wild flowers it does not find favour with the gardener, although it is often encouraged by farmers as it makes good grazing material. By far its greatest fan is the bee, however.

––––––––– *Summer* –––––––––

Salvia pratensis (Meadow clary)

(*Lamiaceae*)	
Good for	Nectar
Type of plant	Perennial
Height	50–80cm
Flowers	Violet blue; June, July

The name clary is thought to be an altered form of 'clear eye' from the time when apothecaries would soak the seeds in water to produce an eye bath to alleviate infections. As with other members of the sage family the flowers are a good source of nectar for bees. This wild *Salvia* would look good in any herbaceous border, so do not just use it in the 'wild' garden.

Onobrychis viciifolia (Sainfoin)
(*Fabaceae*)

Good for	Nectar and pollen
Type of plant	Perennial
Height	30–60cm
Flowers	Pink; June to August

Since its introduction to Britain in the 1650s, sainfoin, a member of the pea family, has been an important agricultural crop – it is cut for hay and the regrowth is used as grazing. It has also escaped into the wild and is recognized in many texts as a wild flower. Whatever its status it is beloved by bees – a good enough reason to grow it.

Daucus carota (Wild carrot)
(*Apiaceae*)

Good for	Nectar and pollen
Type of plant	Biennial
Height	30–60cm
Flowers	One purple or dark red flower at the centre of a mass of white umbels; June to August

Unlike the carrots we grow in the kitchen garden, the wild carrot has a thick, white, woody taproot but this does not matter because in the wild garden we are growing it for its flowers, which are beloved by bees. It has a common country name of 'bird's nest' because as the head of seed develops it becomes concave, giving the whole structure a distinctive 'nest' appearance.

Phacelia tanacetifolia (page 101)
(*Hydrophyllaceae*)

Good for	Nectar and pollen
Type of plant	Annual
Height	Up to 1m
Flowers	Violet blue; June to August

Although not strictly a wild flower, it is often found in the wild as an escapee from field margins that have been set aside for beneficial insects or from game cover crops. It is often used as a green manure to enrich the soil, but in this case it is rarely allowed to flower since at this stage the stems become woody. The nectar-rich flowers open in sequence, giving a long flowering period – ideal for bees.

Summer/autumn

Melilotus officinalis (Melilot or sweet clover) (page 103)

(*Fabaceae*)

Good for	Nectar and pollen
Type of plant	Biennial
Height	Up to 1.2m
Flowers	Yellow; June to September

Although one of its common names is sweet clover, it does not belong to the clover family at all: it is a member of the pea family and looks more like a vetch. It was once commonly grown as a fodder crop and it is said that Gruyère cheese partly owes its flavour to melilot, which grows freely in the Gruyère valley in Switzerland. When it is dried it smells like newly mown hay.

Reseda lutea (Wild mignonette)

(*Resedaceae*)

Good for	Nectar and pollen
Type of plant	Biennial/short-lived perennial
Height	75cm
Flowers	Pale yellow; June to September

Wild mignonette is related to weld (*Reseda luteola*), which has been used for millennia as a dye plant, and common mignonette (*Reseda odorata*), which is grown for its sweet scent. Arguably, wild mignonette's only claim to fame is that it is loved by bees. Reason enough to grow it.

Echium vulgare (Viper's bugloss)
(*Boraginaceae*)

Good for	Nectar and pollen
Type of plant	Biennial
Height	30–90cm
Flowers	Blue-purple with pinkish buds; June to September/October

Do not be put off by its common name! The 'viper' part comes from the fact that 17th-century herbalists thought the speckled stems resembled snake's skin, and 'bugloss' comes from the Greek for 'ox tongue', which is what the leaves look like. It is a striking plant which can grow up to a metre and would not look out of place in the herbaceous border.

Geranium pratense (Meadow cranesbill or meadow geranium)
(*Geraniaceae*)

Good for	Nectar
Type of plant	Perennial
Height	Up to 1m
Flowers	Violet blue, sometimes white; June to September/October

The 'cranesbill' part of the name comes from the fact that after pollination the seed pods look like the bill of a crane. It really is a stunning plant and has given rise to many cultivated varieties, but the original would grace any herbaceous border.

Papaver rhoeas (Common, field or corn poppy) (page 101)
(*Papaveraceae*)

Good for	Pollen
Type of plant	Annual
Height	Up to 60cm
Flowers	Red; June to October

The poppy has long been held as symbol of new life, particularly in agrarian cultures, because of the seed's ability to lie dormant in the earth for many years until the soil is once again cultivated. It was once a common sight in cornfields in Britain, but advances in herbicide technology have all but eradicated it from the crops. Poppies are slowly making a comeback, however, due to farming conservation and stewardship schemes.

In spite of their bright colour they are a useful source of pollen for our buzzy friends – they can detect the pollen because the centre of the flower is well within their visual spectrum. Although each flower only lasts one day, they are produced in large numbers.

Knautia arvensis (Field scabious)

(*Dipsacaceae*)	
Good for	Nectar and pollen
Type of plant	Perennial
Height	Up to 1m
Flowers	Lilac blue; July to September

The flowers of this lovely plant look like miniature pincushions and are a beacon for bees. Its 'cousin', *Knautia macedonica*, with its dark crimson flowers, frequently finds its way into the cultivated garden, but to my mind the subtle colour of *Knautia arvensis* is every bit as good.

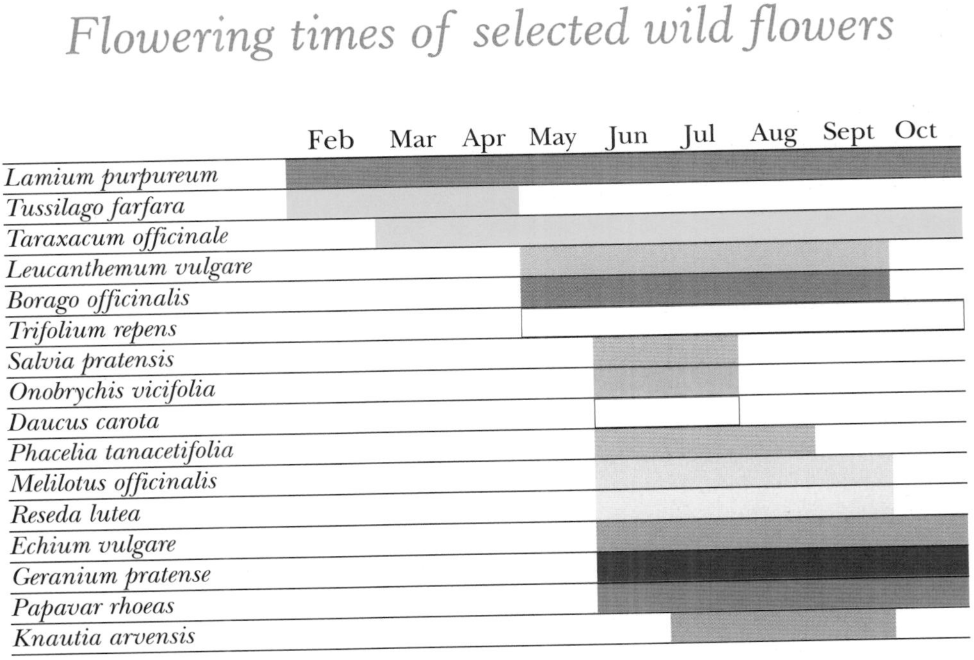

Flowering times of selected wild flowers

Figure 17 Flowering times of selected wild flowers

217

6
Planting Plans

In this chapter I hope you will find inspiration for how to use some of the information that you've gathered so far. To help with this, I have designed five planting plans which fall into two groups. The first group, a small culinary herb bed and two kitchen gardens (or, to give them their posh name, *potagers*), provides mainly food for both humans and bees. The second group, a '3 seasons' mixed border and a wild flower 'meadow', provides a literal feast for bees and a feast for the eyes for humans.

Rather than trying to put too much information on each plan, I have designated areas with letters and/or numbers. You will find the corresponding letter and/or number in the text that accompanies each of the plans so you can see what suggestions for planting I have made. I have also indicated the scale.

I hope that you will find the plans helpful and either use them as they stand or draw inspiration from them and adapt them to suit your own site.

Culinary herb garden

As well as wild flowers, herbs are probably one of the best groups of plants for bees. Even Virgil, writing in 29 BC about bees and their hives, exhorts us to: 'Let green rosemary, and wild thyme with far-flung fragrance, and a wealth of strongly-scented savory, flower around them …' And I think if someone with the means to provide only a window boxful of plants for bees asked me what I would grow the answer would be 'herbs'. (Not that I put myself in the same exalted realm as Virgil, of course!) Not only are they good for bees, but they are also useful in the kitchen, so it is a win–win situation.

The herb garden design that I have created here contains a modest number of varieties which can be used for culinary purposes. There are some obvious omissions, like parsley, but the ones that are included are beloved by bees if, of course, you let some of them flower. This is anathema to a number of herb aficionados, who would argue that the flavour of the herb is lost if you do let them flower, but the flowers of all the herbs in this design are safe for humans to eat (even the bay flowers as long as they are not consumed in vast quantities) so why not experiment with a few in a salad or sprinkle as a garnish?

All the herbs in the design are perennials, which 'regrow' year on year. Even so, they do have a limited lifespan: after a few years they will become woody, even with regular cutting back after flowering, so it is best to replace them periodically.

With a couple of exceptions the herbs in this design do not mind if the soil is a little on the poor side. The exceptions are bay, chives, mint and lemon balm. The bay will appreciate a little organic fertilizer periodically, and the chives and mint would not say no to soil with a more moisture-retentive quality. The preference of the last two is already catered for in the design because I would suggest planting them in a bottomless bucket or other container before plunging the whole thing in the soil so that the rim of the container shows some 2 or 3 centimetres above the surface. If you plant the mint and lemon balm straight into the soil, they will, in no time at all, spread all over the plot, smothering the rest of your precious herbs: they have to be kept in check. And on no account must any of the herbs have their feet in water – they can withstand several degrees of frost, but to have wet feet will spell disaster.

Planting plan for herb garden

Full details of the herbs can be found in the gazetteer.
Newly planted, the plants may appear a little sparse but as they grow and fill the space they will cover the ground so that little soil will be visible.

No. of plants

Bed 1   *Thymus vulgaris*
(common or garden thyme)   4

Bed 2   *Melissa officinalis* (lemon balm)   1
surrounded by *Salvia officinalis* 'Tricolor'
(tricolour sage)   6

Figure 18 Plan of herb garden

| Bed 3 | *Thymus polytrichus subsp. britannicus* (wild creeping thyme) | 4 |

| Bed 4 | *Mentha sativa* (mint) | 1 |
| | surrounded by *Satureja montana* (winter savory) | 6 |

| Bed 5 | Standard ('lollipop' shape) *Laurus nobilis* (bay) | 1 |
| | underplanted with *Rosmarinus officinalis* (*Prostratus* group) 'Capri' (prostrate rosemary) | 2 |

| Bed 6 | *Mentha suaveolens* (apple mint) | 1 |
| | surrounded by *Hyssopus officinalis* (hyssop) | 6 |

| Bed 7 | *Thymus pulegiodes* (broad-leaved thyme) | 4 |

| Bed 8 | *Melissa officinalis* 'Aurea' (variegated lemon balm) | 1 |
| | surrounded by *Origanum vulgare* (marjoram) | 6 |

| Bed 9 | *Thymus* 'Hartington Silver' (Hartington thyme) | 4 |

Beds 2, 4, 6 and 8		
	are circled by *Lavandula angustifolia* 'Hidcote' (Hidcote lavender)	12
	and divided by *Allium schoenoprasum* (chives)	12

Potager

The word potager comes from the French *potagère*, meaning vegetable garden, but the term potager has come to mean something a little more ornamental than just an old-fashioned veg plot. Yes, it contains vegetables, fruit and herbs, but flowers are also allowed within its bounds and it is all planted out in such a way as to be aesthetically pleasing as well as functional – William Morris would have approved, I think.

I have provided plans for two potagers. The first plan covers a large area, some 16 by 16 metres: an area, I admit, which is beyond what most people have at their disposal, but the principle of planting is the same whatever size your plot, so you can scale down (or up) accordingly.

The second plan is much more modest – 7 by 6 metres – and I have included it because there is a limit to how much scaling down of the large plan you can do without it becoming impractical. It still includes most of the elements of the large plan, but is better suited to a smaller space.

You will note that I have not included specific cultivation notes. These are beyond the scope of this book, and there are numerous excellent guides to growing vegetables, fruit and flowers for cutting, available from good booksellers.

There are some permanent 'structure' and perennial plants (fruit trees and bushes) and space for herbs detailed on both plans. The larger one also includes areas of flowers for cutting. I have arranged the rest of the plot, however, in groups of vegetables.

Vegetables

The groups of vegetables are ordered in such a way that it should be fairly easy to keep to a rotational planting regime, whereby the same type of vegetable is not grown in the same place each year – in fact it should be five years before your beans are back to where they started. So in Year 1 you would plant up according to the plan; in Year 2 you shift things round in a clockwise direction so that Bed 1 is now planted up with what was in Bed 2 the previous year, and so on. The main reason for rotating crops is to prevent a build-up of diseases and pests which target particular types of plants. If you do not grow the same thing in the same place year on year then the disease has less chance to get a foothold. Another reason is that some vegetables require more nutrients than others. The legume family 'fix' nitrogen in the soil, a nutrient for which the brassica family is hungry, so it makes sense to follow peas and beans with cabbage or broccoli.

You will see that I have not detailed precisely what vegetables to plant. There is no point in suggesting you grow parsnips if no one in your family likes them! So I have kept to ideas for vegetables within each 'group' so that you can pick and choose what you would like to plant.

I have also included vegetables which are not particularly useful to bees (such as peas, which are self-fertile), but we like them and the bees are well catered for in the rest of the potager.

If you wish to keep seed from your favourites then you must obviously allow some of the vegetables to flower – even the broccoli, onions and carrots! This is where the bees come in, of course. A note of caution, though: if you grow F1 hybrids they will not come true from seed so if you want to collect your own seed, make sure you grow 'traditional' varieties. Varieties marked 'heirloom' are also worth growing because they are often closer to the original species than modern hybrids.

Year 1

Bed 1 **Legumes** – beans (runner, French and broad), peas (including sugar snap and mangetout)

Bed 2 **Brassicas** – cabbage, Brussels sprouts, broccoli, swede, turnip, kohlrabi (although the last three are generally classified as roots, they have been included here because they are of the cabbage 'family' and are prone to the same diseases)

Bed 3 **Onions and others** – onions, garlic, leeks, courgettes, pumpkins

Bed 4 **Roots** – potatoes, carrots, parsnips, celeriac, outdoor tomatoes (tomatoes are included here because they are the same family as potatoes)

Year 2

Bed 1	Brassicas
Bed 2	Onions and others
Bed 3	Roots
Bed 4	Legumes

and so on …

There are some vegetables that can be slotted in anywhere there is space because they hold no threat of disease to other vegetables. These include: leafy salad crops such as lettuce and rocket, also spinach, Florence fennel (this is the bulb variety as opposed to the herb grown for its leaves) and chard.

Fruit

Fruit trees, bushes and plants are of inestimable value to both bees and humans. From a bee's point of view there are few months during the foraging season when there is not some sort of fruit in flower, ranging from apples in spring to blackberries, which often have a second flush of flowers even when there are berries ripening on the branch. And, from a human perspective, fruit provides a vital source of nutrition, so it really is a 'win–win' group of plants. I have included a range of fruit in the potager, none of which needs to be included in a rotational planting scheme. Once planted, they can stay where they are, or, like strawberries, be renewed as necessary. Again, I have not detailed any specific varieties, and there are a couple of reasons for this. First, personal preference – just because I like a particular variety of gooseberry, say, it does not necessarily follow that everyone else will. Second, and perhaps more importantly, geographical location – what will grow well in the south-east of England may not do so well in the north-west, so before you decide on a variety of apple, for example, consult a specialist grower to see what will not only survive in your part of the country, but also thrive.

You will see that I have suggested that the apples and pears are trained as 'step-over' trees. These are trees on a dwarfing rootstock that have a very short centre stem of about 30cm – low enough to step over, hence the name – with a horizontal branch either side. They are both attractive and productive.

Herbs

The centre of the large potager shows the culinary herb design featured in this chapter surrounded by alpine strawberries. Other herbs such as parsley, coriander, chervil and French tarragon are also included elsewhere. Herbs are also included in the small plan: they have their own modest beds and are also used as underplanting.

Flowers

Some people would argue that the vegetables themselves and their own flowers provide interest and colour enough without introducing any extra blooms. It is true that the crinkly, dark leaves of the savoy cabbage, the glaucous-blue sword-like foliage of leeks, or the brilliant scarlet and yellow stems of chard have an attraction all their own. And the flowers of some vegetables are truly beautiful: think of the scarlet or even orange flower of the runner bean (which was grown for its flowers long before it dawned on anyone to eat the beans!), the crimson-flowered broad bean, or the flamboyant flower of the courgette (page 105) (which, incidentally you can eat). And if you accidentally-on-purpose let some of your other vegetables flower you can enjoy the tracery globe of onions and leeks, the beautiful white umbels of carrots, and the purple, daisy-like flowers of *Tragopogon porrifolius* (salsify).

I always try to make room for 'proper' flowers in my kitchen garden, however. They will attract no end of beneficial insects – and especially bees – into the potager to help pollinate the vegetable crops but they will also provide a source of beautiful cut flowers for the house. And you do not have to be a flower arranger to make an attractive display – the flowers do that for you (Figure 88)!

Green manure

In addition, it is advisable not to leave soil bare, either during the growing season or over winter. Weeds will soon take hold, or the nutrients will be leached out by rain, so why not plant some 'green manure' – plants which can be dug into the soil to increase the fertility and add humus.

Green manure includes *Trifolium* (clover), *Phacelia*, *Medicago sativa* (alfalfa or lucerne) and *Melilotus officinalis* (sweet clover or melilot), all of which, if left to flower, are beacons for bees. If you do not want them self-seeding everywhere then be sure to dig them in immediately after flowering.

A note, too, about the paths. The main paths in the large potager are wide enough to take a wheelbarrow and the minor paths in the large potager and all the paths in the smaller one are still of sufficient width to walk along comfortably. What the paths are made of will chiefly depend on preference and/or how much you can afford to spend on them. Two of the cheapest options are to either grass them, which will require mowing and the grass can get worn with use and very soggy if it rains a lot, or to compact the bare soil, cover it with a permeable membrane and spread it with a layer of bark chippings. The latter I have found is the least desirable because the chippings end up at the side of the path, exposing the membrane, which can become very slippery.

Another option is to use gravel, but this can be expensive. Alternatively you can use self-binding gravel (examples of self-binding gravel are hoggin and Breedon gravel). Self-binding gravel is a mixture of gravel and 'sticky' sand which naturally holds together and sets hard when it is wetted and compacted. This surface is my favourite in the kitchen garden: it is hard-wearing, 'solid', is environmentally sound (it is usually quarried and distributed over a local area and often travels no more than 50 kilometres from source) and, aesthetically speaking, is very attractive.

A basketful of bee-friendly cut flowers

Planting plan for large potager

The plan is set out geometrically, partly for aesthetic reasons but mainly because it is easier to access and manage.

Herbs This is the separate plan for culinary herbs, given elsewhere in this chapter. In addition, it is surrounded by alpine strawberries.

BH – Beehive If you welcome a beehive in your potager, then a good place to put it is in the corner amongst the gooseberries. The bees will be less disturbed here. You can site your beehive(s) anywhere, of course (pages 231 and 233). You will see in both cases that the hives have been placed on stands or a wooden pallet. This is to keep them off the ground to aid ventilation and to prevent dampness from penetrating.

Beds 1 to 4 are described above

A1 and A2 – 'Step-over' **apples** underplanted with **parsley**

A3 and A4 – 'Step-over' **apples** underplanted with **French tarragon**

P1 and P2 – 'Step-over' **pears** underplanted with **coriander**

P3 and P4 – 'Step-over' **pears** underplanted with **chervil**

F1, F4, F5 and F8 – Perennial flowers for cutting, including *Achillea, Polemonium, Eryngium, Verbena, Echinops, Sedum, Perovskia, Liatris, Echinacea, Aster, Rudbeckia, Dahlia*. (See gazetteer for details)

F2, F3, F6 and F7 – Annual flowers for cutting, including *Iberis, Clarkia, Centaurea, Cosmos, Nigella, Helianthus*. (See gazetteer for details)

G – Gooseberries

B – Blackberries

R – Raspberries

C – Currants, black and red

L – Lavender

RY – Rosemary

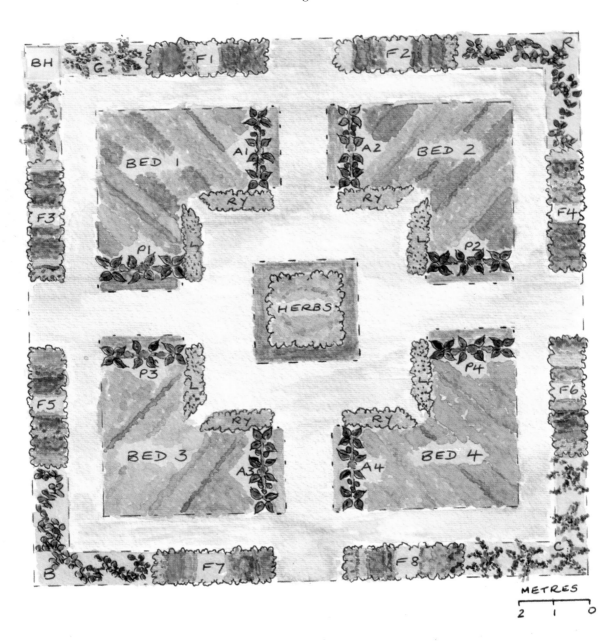

Figure 19 Plan of large potager

Planting plan for small potager

BH – Beehive The size of this potager is too small to solely sustain a hive of bees, but bees will fly up to three miles to find food, so what lies beyond the perimeter will fall within their foraging area.

H – Herbs There are three small areas set aside for herbs. Rather than prescribe what you should grow in them, I have left you to choose your favourites or the ones that you are most likely to use in the kitchen. For example, I cannot do without tarragon, marjoram, parsley, thyme and chives, so they would be at the top of my list.

R – Raspberries Or they could equally be blackberries!

A – Espalier apples underplanted with **strawberries**, both garden and alpine

C - Currants, black and red I like currants but there is no reason why you cannot have gooseberries instead.

Beds 1 to 4 are described above

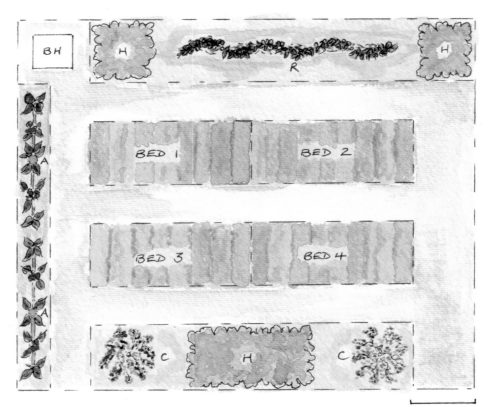

Figure 20 Plan of small potager

I METRE

Beehives in a potager

3 seasons border

This planting plan is for what you might call a 'traditional' border which aims to offer flowers for bees throughout their foraging period (spring to autumn) – that's why it is not a '4 seasons' border!

I have said elsewhere that it is much better to have blocks of plants rather than single specimens: this is far easier to achieve if you have a big garden, of course. In this plan you will find single specimens of shrubs but at least three of each perennial or annual plant grouped together – this is the best we can hope for given the area we are dealing with. If you do have more space then you can adjust the plan by increasing the number of plants in each group and even include a tree. If you have less space, however, then you will either have to decrease the number of plants in each group or judiciously leave out one or more groups altogether: given that I have included more than one suggestion for each season, the latter should be fairly easy to achieve.

The plan is built up in 'layers' of plants. First, there are some shrubs which give the 'backbone' to the design: these are permanent and will give structure as well as year-round interest.

Second, there are perennial, herbaceous plants which form the main planting group: they are nearly all deciduous (although in a mild winter a few may keep some leaves) but all will come back year after year.

Third, there are the temporary, annual or tender perennial plants which can be used to fill the gaps. These are not marked on the plan itself but given as 'additional extras' in the planting schedule. As the border matures, you may find that the gaps become fewer and you will not need as many 'fillers'.

Last, but to my mind as important as any of the other groups, are the bulbs: these are an invaluable group for providing colour – and food for bees – during the spring especially and, if you distribute them throughout the border, they give coherence to the overall design.

You will also notice that I have chosen quite a restricted colour scheme, limiting myself to, in the main, only two groups of colours – yellows and blues/lavenders/purples. There are two reasons for doing this: first, because bees can easily detect these shades and hues, and second, because yellow and purple are complementary colours and, from a design point of view, are bound to look good together.

Beehive (and Toady) in a potager

I have assumed that the border has a 'backdrop' of some sort – an existing wall, fence or hedge. If you wish to position it as an island bed, then you will have to adjust some of the planting a little as, with the design as it stands, most of the taller species tend to be at the back so that they do not obscure the shorter ones. The clematis, which is to be found against the wall or fence, could be left out altogether or grown up a slim, pyramid-shaped trellis positioned in the middle of the bed.

I have also assumed that the soil is neutral, moist but free-draining and that the site is in either full sun or is in sun for at least part of the day.

The plan indicates how much space the plants will take up two to three years after planting so do not be alarmed if, when they are newly planted, they appear a little sparse: this is when you can supplement the planting with more annuals.

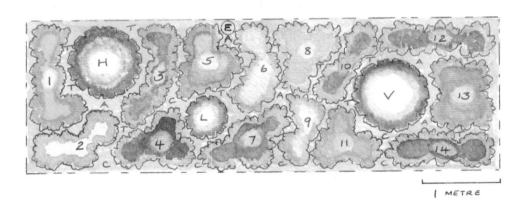

Figure 21 Plan and illustration of 3 seasons border

Planting plan for 3 seasons border

———————— Shrubs and climber ————————

H	*Hydrangea aspera* 'Macrophylla'	1
L	*Leptospermum scoparium*	1
V	*Viburnum tinus* 'Eve Price'	1
E	*Clematis viticella* 'Etoile Violette'	1

———————— Perennials ————————

1	*Kniphofia* 'Bee's Lemon'	3
2	*Lamium maculatum* 'White Nancy'	3
3	*Polemonium caeruleum*	3
4	*Lavandula angustifolia* 'Hidcote'	3
5	*Doronicum orientale* 'Magnificum'	3
6	*Papaver orientale* 'Wedding Day'	3
7	*Aster amellus* 'Veilchenkönigin'	3
8	*Achillea* 'Credo'	3
9	*Origanum vulgare*	3
10	*Perovskia atriplicifolia* 'Blue Spire'	3
11	*Echinacea* 'Sunrise'	3
12	*Echinops ritro* 'Veitch's Blue'	3
13	*Helenium* 'El Dorado'	3
14	*Geranium* 'Johnson's Blue'	3

———————— 'Fillers' ————————

Bulbs

| A | *Allium* 'Purple Sensation' | 20 in total in 5 groups |
| C | *Crocus chrysanthus* 'Cream Beauty', 'Blue Pearl' and 'Snow Bunting' and *Crocus vernus* 'Remembrance' and 'Yellow Mammoth' | 60 in total in 6 groups |

T   *Tulipa* – a mixture of some of the   60 in total in 10 groups
          following varieties: 'Purissima', 'Candela',
          'Sistula', 'Spring Green', 'West Point',
          'Alabaster' and 'City of Vancouver'

Annuals, biennials and tender perennials

A mixture of some or all of the following, depending on how many gaps there are: *Cheiranthus cheiri* (yellow variety), *Heliotropium arborescens* (page 145), *Limnanthes douglasii* (page 147), *Reseda odorata, Nigella hispanica, Verbena rigida*

Flowering times of plants in 3 seasons border

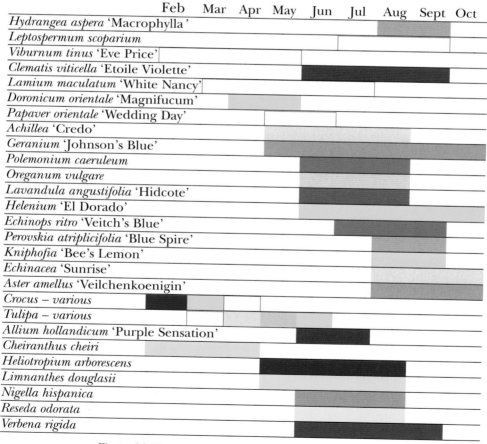

Figure 22 Flowering times of plants in 3 seasons border

Wild flower garden

We know that wild flowers are some of the best 'bee' plants going and a good number of them, to my mind at least, can hold their heads up alongside their more sophisticated, cultivated counterparts – a touch of *My Fair Lady* perhaps, but like Eliza Doolittle they have a charm all of their own.

A planting plan for a wild flower area is probably one of the most difficult plans to put together. By its very nature a wild flower area appears to be random, a bit higgledy-piggledy, so to try and force any sort of design or structure on it seems somewhat futile. However, the chances are that even what look like completely 'natural' areas of wild flowers within an otherwise managed landscape have been carefully prepared and sown, and subsequently looked after. Which is what we must do to create something which is both attractive to us and, more importantly, to the bees.

As with many other jobs, preparation is the key. All of the wild flowers which are of benefit to bees that I have suggested for this plan require more or less the same conditions, namely, a clean, prepared bed of neutral, loamy soil which is not too fertile, and full sun or sun for a good part of the day. (There are bee-friendly flowers that grow in the shade, in calcareous or acidic soil but for the purposes of this plan I am keeping to the 'best option' so that even if I cannot guarantee total success, I can at least say that it will not be a complete failure!) The most important thing is to make sure that the area you want to plant is completely free of weeds, both annual and perennial.

If the area you want to earmark for wild flowers has had things growing on it before, bear in mind how fertile the soil might be. Generally speaking wild flowers grow best in soil that has not been enriched with lots of organic matter. Think about what their natural habitat is: wild flowers mostly grow in grassland, wasteland, road verges or other rough land which has not been cultivated or improved, so we need to try and recreate those sorts of conditions. If the soil is too 'rich' then it is worth removing a couple of inches and using it elsewhere in the garden.

Once you have removed all the weeds, dig the soil over, firm it and lightly rake it. Now you can either sow seeds (by far the cheapest option) or plant plug plants that you can buy from reputable seedsmen and growers. Although it may sound perverse to do so, if you are using seeds, then sow them in straight lines in drills that you have already watered and then *lightly* cover them over with soil. 'Straight lines?' I hear you cry. Yes, sowing them like this will make it much easier to keep on top of unwanted intruders until 'your' seeds have germinated and put some roots down – you can simply hoe between the rows. You may need to thin out the seedlings. If they are too dense then each little plant will be competing against its neighbour and may become thin and straggly as a result. Once the plants have grown and filled out the space, you will never be able to tell that they started out like a company of soldiers. Both seeds and plug plants will need to be watered until they become established but thereafter they should be able to survive with what nature provides.

The wild flower patch will have to be cared for. Spent flowers must be removed after they have seeded and because there is a mixture of spring and summer flowering plants this should be delayed until the early autumn. Take off all the dead material down to about 7.5cm and leave it to dry off for a few days then rake it up. The main reason for not leaving any dead vegetation is that it would rot down, smothering the crowns of any perennial plants and adding fertility to the soil, which we do not want. Next spring perennial plants will come back to life and the seed shed from annual and biennial species will germinate to start the cycle again.

The area on the planting plan measures 2.4 metres by 0.8 metres – this is a reasonably large area. If you only have a fraction of that area that you can use for wild flowers, then the plan is easy to scale down or, conversely, if you have a much larger area, you can increase the size of each area, or even introduce additional species.

The amount of seed you need for such an area is relatively small. The usual recommended sowing rate for wild flower areas is between 3 and 5 grams per square metre, so you will only need about 10 grams in total for the suggested planting plan. Seed merchants will supply individual species in amounts as little as 1 gram so the 11 different species that I have suggested fit in with this quite nicely.

Each different flower is planted in two blocks separate within the whole area. There are two reasons for this. Firstly, we know that bees are attracted to masses of plants of the same species. Secondly, it is to make sowing the seeds (in straight rows, as explained above) easier. As the plants become established and start to self-seed, the edges of the 'blocks' will become blurred and may even disappear altogether, but by this stage we can let go of the reins a little and let nature take her course.

You will notice that I have not included what can be regarded as one of the 'bee' plants *extraordinaire, Taraxacum officinale* (dandelion). There is a simple reason for this – I guarantee that it will find its own way into your wild flower patch without any human intervention! You may even get to the point where you are digging some up so that they do not take over.

What you could include, but which are not shown on the planting plan, are some bulbs. The obvious choice for a native bulb would be *Hyacinthoides non-scripta* (native bluebell) but this favours woodland conditions and we are working with an open, sunny site. One bulb that does fit our requirements is *Ornithogalum pyrenaicum* (Bath asparagus). This is usually found in the west of England, and its young spikes have been eaten like asparagus – hence its name of Bath asparagus.

Planting plan for wild flower garden

———— List of species ————

1. *Lamium purpureum*, 2. *Tussilago farfara*, 3. *Leucanthemum vulgare*
4. *Trifolium repens*, 5. *Salvia pratensis*, 6. *Phacelia tanacetifolia*
7. *Melilotus officinalis*, 8. *Reseda lutea*, 9. *Echium vulgare*
10. *Geranium pratense*, 11. *Knautia arvensis*

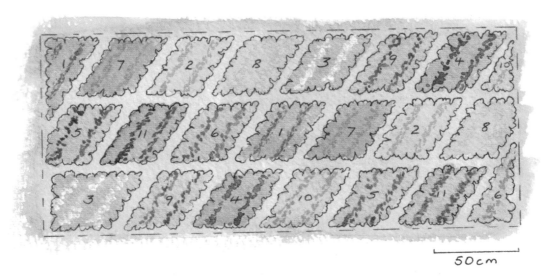

50cm

Figure 23 Plan of wild flower garden

The plants I have chosen will grow just about anywhere in the UK given a neutral soil. If, however, you want to tailor your choice to what will grow best in your particular locality, then the Natural History Museum has compiled an online database which details the wild flowers which occur naturally in each part of the country. It is free to access it and all you need to do is enter the first part of your postcode in the 'search' box. You can find the website details in the 'Further Reading' section near the end of the book.

Flowering times of flowers in wild flower garden

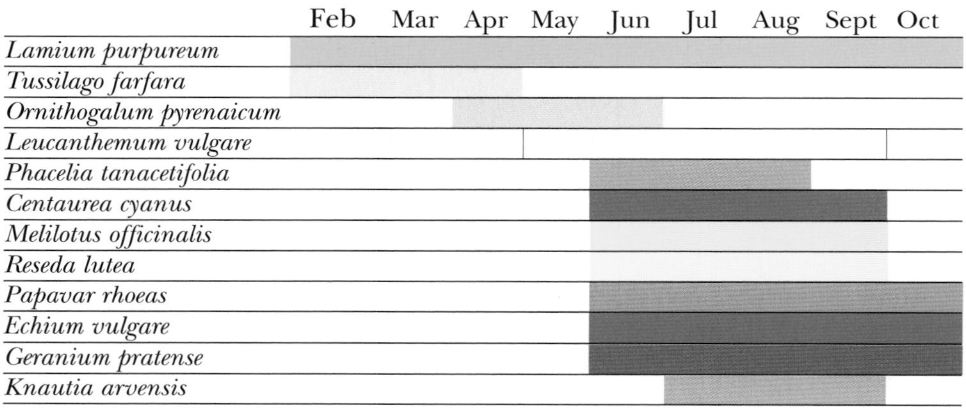

	Feb	Mar	Apr	May	Jun	Jul	Aug	Sept	Oct
Lamium purpureum									
Tussilago farfara									
Ornithogalum pyrenaicum									
Leucanthemum vulgare									
Phacelia tanacetifolia									
Centaurea cyanus									
Melilotus officinalis									
Reseda lutea									
Papavar rhoeas									
Echium vulgare									
Geranium pratense									
Knautia arvensis									

Figure 24 Flowering times of plants in wild flower garden

Appendix 1
Annuals and Biennials of Value to Bees

There are many annuals which are attractive to bees: here are a few worth growing.

Alyssum sp.
Amberboa moschata
Calendula sp.
Callistephus chinensis
Centaurea cyanus (page 39)
Cheiranthus cheiri (page 77)
Clarkia sp.
Cosmos bipinnatus (page 145)
Eschscholzia californica
Godetia sp. (single-flowered)
Helianthus annuus (page 98)
Heliotropium cultivars (page 143)
Iberis sp.
Limnanthes douglasii (page 147)
Myosotis sp.
Nigella damascena
Nigella hispanica
Reseda odorata

Appendix 2
Perennials and Bulbs of Value to Bees

Achillea sp. (page 155)

Agastache sp.

Allium sp.

Anchusa azurea

Anchusa officinalis

Anemone blanda

Anemone x *hybrida*

Arabis sp.

Armeria maritima

Aster sp. (page 82)

Aubretia sp.

Centaurea sp.

Centranthus ruber

Colchicum autumnale (page 112)

Crocus sp.

Cynara cardunculus

Dahlia (single-flowered varieties) (page 111)

Doronicum sp.

Echinacea sp. (page 82)

Echinops sp. (page 94)

Eranthis hyemalis

Erigeron sp.

Eryngium sp. (page 94)

Gaillardia sp. (page 163)

Geranium sp. (page 76)

Geum sp.

Helenium sp. (page 81)

Helleborus sp.

Kniphofia sp. (page 166)

Lamium sp. (page 211)

Leucanthemum sp.

Liatris spicata

Muscari sp.

Myosotis sp.

Nepeta sp. (page 7)

Papaver orientale

Penstemon sp. – some with large corollas (page 36)

Perovskia atriplicifolia (page 89)

Persicaria amplexicaulis

Polemonium caeruleum (page 76)

Pulmonaria sp.

Rudbeckia sp. (page 31)

Salvia sp.

Scabiosa sp.

Sedum spectabile (page 81)

Sidalcea malviflora

Solidago sp.

Stachys byzantina (page 95)

Tulipa sp. (page 74–5)

Verbena bonariensis

Verbena rigida (page 59)

Veronica longifolia

Veronicastrum sp.

Appendix 3
Shrubs and Climbers of Value to Bees

Abelia chinensis
Abelia schumanii
Arbutus unedo
Berberis sp.
Buddleja davidii (page 85)
Buxus sempervirens
Calluna vulgaris
Caryopteris sp.
Ceanothus sp.
Chaenomeles speciosa
Chaenomeles x *superba*
Clematis sp.
Cotoneaster horizontalis
Cytisus sp.
Daphne mezereum
Enkianthus campanulatus
Erica sp.
Escallonia bifida
Eucryphia sp. (page 109)
Fuchsia sp. (page 108)
Hebe sp.

Hedera helix (page 110)
Helianthemum sp.
Hydrangea aspera 'Macrophylla' (page 9)
Lavandula sp. (page 88)
Leptospermum scoparium
Mahonia aquifolium
Olearia x *haastii*
Perovskia atriplicifolia (page 89)
Philadelphus sp.
Potentilla fruticosa
Prunus laurocerasus (page 176)
Romneya coulteri
Rosa sp. (single-flowered)
Rosmarinus officinalis
Sarcococca sp.
Skimmia japonica
Symphoricarpos albus
Ulex minor
Viburnum tinus (page 73)

Appendix 4
Trees of Value to Bees

Acer pseudoplatanus

Aesculus hippocastanum

Castanea sativa

Catalpa bignonioides (page 87)

Cercis siliquastrum

Corylus avellana (page 84)

Crataegus monogyna

Ilex aquifolium

Liriodendron tulipifera

Malus sylvestris

Prunus amygdalus

Prunus armeniaca

Prunus avium

Prunus cerasus

Prunus persica

Prunus spinosa

Rhamnus catharticus

Robinia pseudoacacia

Salix alba

Salix caprea

Sophora japonica

Sorbus aria

Sorbus aucuparia

Tetradium daniellii

Tilia sp.

Appendix 5
Edible Plants of Value to Bees

Having a separate section of 'edibles' of value to bees may seem a little odd, because when you think about it any plant that produces a fruit needs to be pollinated. I have included this list, however, because as well as containing the obvious suspects like apples, strawberries, beans and squashes, it also includes some which are a little less obvious, but will still be attractive to bees. This is particularly true of those vegetables which we normally harvest before they flower, like radishes and onions. If you let these produce flowers (which in vegetable circles is known as 'bolting'), however, you will be surprised at just how many insects they will attract.

You will note that I have broken with my rule of giving Latin names only in this list. The reason for this is that some 'edibles', especially vegetables, are universally known by their common names. However, I have, where appropriate, given the Latin name in case you wish to look up the plant elsewhere in the book.

Fruits

Apple	*Malus domestica*
Blackberry	*Rubus fruticosus* (page 23)
Blackcurrant	*Ribes nigrum*
Cherry	*Prunus avium*
Gooseberry	*Ribes uva-crispa* var. *reclinatum*
Medlar	*Mespilus germanica*
Pear	*Pyrus communis*
Plum	*Prunus domestica*

Raspberry	*Rubus idaeus*
Redcurrant	*Ribes rubrum*
Strawberry	*Fragaria* x *ananassa* (page 21)
Loganberry	*Rubus* x *loganobaccus*
Jostaberry	Various but the Royal Horticultural Society lists it as *Ribes* x *culverwellii*

Nuts
Hazel	*Corylus avellana*
Walnuts	*Juglans* sp.

Vegetables
Asparagus
Broad beans
Broccoli
Cabbage
Calabrese
Carrots
Cauliflower
Chicory
Courgettes
Cucumbers
Florence fennel
French beans
Kale
Leeks
Marrows
Onions
Parsnips
Pumpkins
Radishes
Runner beans
Salsify
Swedes
Turnips

Appendix 6
Herbs of Value to Bees

Herbs are perhaps one of the best groups of plants for bees. Just remember to let them flower.

Allium schoenoprasum
Anethum graveolens
Angelica archangelica
Borago officinalis (page 24)
Calendula officinalis (page 97)
Centaurea cyanus (page 39)
Chicorium intybus
Echinacea sp. (page 82)
Eupatorium purpureum
Foeniculum vulgare
Hesperis matronalis
Hyssopus officinalis (page 67)
Laurus nobilis
Lavandula sp. (page 88)

Melissa officinalis
Mentha sp. (page 68)
Myrrhis odorata
Nepeta sp. (page 7)
Oenothera biennis
Origanum vulgare (page 25)
Rosmarinus officinalis
Salvia officinalis
Saponaria officinalis
Satureja hortensis
Satureja montana
Stachys officinalis
Symphytum officinale
Tanacetum parthenium
Tanacetum vulgare
Thymus vulgaris
Tropaeolum majus

Appendix 7
Wild Flowers of Value to Bees

Achillea millefolium
Borago officinalis (page 24)
Bryonia dioica
Caltha palustris
Centaurea cyanus (page 39)
Centaurea scabiosa
Chamerion angustifolium
Cirsium sp.
Clematis vitalba
Daucus carota
Dipsacus fullonum
Echium vulgare
Eupatorium cannabinum
Filipendula ulmaria
Galium verum
Geranium pratense
Impatiens glandulifera
Knautia arvensis
Lamium purpureum
Leucanthemum vulgare (page 37)

Linaria vulgaris
Lotus corniculatus
Lythrum salicaria
Malva sp.
Melilotus officinalis (page 103)
Onobrychis viciifolia
Ornithogalum pyrenaicum
Papaver rhoeas
Phacelia tanacetifolia (page 101)
Ranunculus acris
Ranunculus ficaria
Reseda lutea
Salvia pratensis
Stachys officinalis
Succisa pratensis
Taraxacum officinale
Trifolium dubium
Trifolium repens (page 25)
Tussilago farfara
Valeriana officinalis

Further Reading

Aston, D. and Bucknall, S. (2004) *Plants and Honeybees: An Introduction to Their Relationships.* Northern Bee Books.

Brickell, C. (ed.) (1989) *The Royal Horticultural Society Gardener's Encyclopedia of Plants and Flowers.* Dorling Kindersley.

Brickell, C. (ed.) (2007) *The Royal Horticultural Society Encyclopedia of Gardening.* Dorling Kindersley.

Brickell, C. and Joyce, D. (2006) *RHS Pruning and Training.* Dorling Kindersley.

Brown, Deni (2003) *The Royal Horticultural Society Encyclopedia of Herbs and Their Uses.* Dorling Kindersley.

Cramp, D. (2008) *A Practical Manual of Beekeeping.* Spring Hill.

De Bruyn, C. (1997) *Practical Beekeeping.* The Crowood Press Ltd.

Feltwell, J. (2006) *Bumblebees.* Wildlife Matters.

Foster, A.M. (1988) *Bee Boles and Bee Houses.* Shire Publications Ltd.

Hamilton, Geoff (1997) *The Organic Garden Book.* Dorling Kindersley.

Hooper, T. and Taylor, M. (2006) *The Bee-Friendly Garden.* Alphabet and Image Publishers.

Howes, F.N. (1945) *Plants and Beekeeping*. Faber and Faber. Available as a download from www.archive.org

Hughes, C. (2010) *Urban Beekeeping*. The Good Life Press.

International Bee Research Association (2008) *Garden Plants Valuable to Bees*. IBRA.

Jill, Duchess of Hamilton, Hart, P. and Simmons, J. (2000) *English Plants for Your Garden*. Frances Lincoln.

Kirk, William (2006) *A Colour Guide to Pollen Loads of the Honeybee*. International Bee Research Association.

Kollerstrom, N. (yearly) *Gardening and Planting by the Moon*. Quantum.

Larkcom, J. (2002) *Grow Your Own Vegetables*. Frances Lincoln.

McVicar, J. (2009) *Jekka's Complete Herb Book*. Kyle Cathie.

Mountain, M.F. (1965) *Trees and Shrubs Valuable to Bees*. International Bee Research Association.

Osler, Mirabel (2000) *A Gentle Plea for Chaos*. Bloomsbury Publishing.

Pears, Pauline and Stickland, Sue (1999) *Royal Horticultural Society Organic Gardening*. Mitchell Beazley.

Raven, S. (1996) *The Cutting Garden: Growing and Arranging Garden Flowers*. Frances Lincoln.

Royal Horticultural Society (2008) *Vegetable and Fruit Gardening*. Dorling Kindersley.

Royal Horticultural Society (2010) *Royal Horticultural Society Entomology Advisory Leaflet No 6204*.

Useful Addresses and Websites

Plant related

Royal Horticultural Society
80 Vincent Square
London SW1P 2PE
www.rhs.org.uk

Garden Organic
Coventry
Warwickshire CV8 3LG
United Kingdom
www.gardenorganic.org.uk

Perennials by Design
151a Southport New Road
Tarleton
Preston
Lancashire PR4 6HX
Contact Tricia Brown 01772 812672
Specialist perennial nursery, open to
the public

Cotswold Garden Flowers
Sands Lane
Badsey
Worcestershire WR11 7EZ
www.cgf.net
Specialist nursery, open to the public.
Mail order sales

Keepers Nursery
Gallants Court
East Farleigh, Maidstone
Kent ME15 0LE
www.keepers-nursery.co.uk
Supplier of fruit trees

Chew Valley Trees
Winford Road
Chew Magna
Bristol BS40 8HJ
www.chewvalleytrees.co.uk
Trees and hedging

Trees Please
Dilston Haugh Farm
Corbridge
Northumberland NE45 5QY
www.treesplease.co.uk
Trees and hedging

Jekka's Herb Farm
Rose Cottage
Shellards Lane
Alveston
Bristol BS35 3SY
www.jekkasherbfarm.com
Organic herb grower. Mail order and
online sales

Emorsgate Wild Seeds
Limes Farm
Tilney
All Saints
King's Lynn
Norfolk PE34 4RT
www.wildseed.co.uk
Supplier of wild flower seeds

British Wildflower Plants
Burlingham Gardens
Main Road
North Burlingham
Norwich NR13 4TA
www.wildflowers.co.uk
Supplier of wild flower plants

David Austin Roses
Bowling Green Lane
Albrighton
Wolverhampton WV7 3HB
www.davidaustinroses.com
Supplier of roses of all descriptions

Natural History Museum
Cromwell Road
London SW7 5BD
www.nhm.ac.uk/nature-
online/life/plants-fungi/postcode-plants
Free online database of wild flowers
that can be found in your locality

The Real Cut Flower Garden
Walkers Cottage
Clifford
Hereford HR3 5HQ
www.cutflowergarden.co.uk
Seasonal cut flower grower

Bee related

**International Bee Research
Association** (IBRA)
18 North Road
Cardiff CF1 3DY
www.ibra.org.uk

British Beekeepers' Association
National Beekeeping Centre
National Agricultural Centre
Stoneleigh
Warwickshire CV8 2LZ
www.britishbee.org.uk

Bumblebee Conservation Trust
School of Biological and
Environmental Sciences
University of Stirling
Stirling FK9 4LA
www.bumblebeeconservation.org.uk

Crossmoor Honey Farm
Keeps Barn Farm
Crossmoor
Preston
Lancashire PR4 3XB
www.crossmoorhoney.com
Bees, equipment and honey

Glossary

Bee things

Bee bread
A mixture of pollen, honey, hypopharyngeal secretions and enzymes which is stored in the comb.

Brood
The offspring of a queen.

Brood food
See 'Royal jelly'

Comb (honeycomb)
Wax built into hexagonal cells which are used for rearing brood, and storing pollen and honey.

Corbicula
A slightly concave indent, fringed with hairs, on the outer surface of each hind leg of the worker bee.

Drone bee
A male bee.

Hypopharyngeal glands
Glands located in the head of worker bees which produce, at various times, royal jelly and enzymes.

Pollen basket
See '*Corbicula*'

Proboscis
The feeding structure of a bee, consisting of several mouth parts that are brought together to form a means of sucking water or nectar.

Queen bee
A fertile female bee.

Royal jelly (brood food)
A secretion from the hypopharyngeal glands which is fed to larvae and the queen.

Swarm
A group of bees led by a queen that leaves the parent hive in order to start a new colony.

Varroa mite
A parasitic mite of honeybees.

Worker bee
A sterile female bee.

Plant things

Annual plant
A plant that completes its life cycle in one year.

Anther
The part of the stamen that contains pollen in pollen sacs.

Biennial plant
A plant that completes its life cycle over a two-year period.

Bulb
An underground storage organ which consists of a stem and bud surrounded by fleshy leaves.

Corm
A short vertical underground stem that acts as a storage organ for food.

Corolla
All the petals of a flower. Sometimes these are 'fused' to form a tube.

Cultivar
A plant variety produced in cultivation, often by selective breeding. A cultivar is indicated by enclosing its name in single quotation marks, for example, *Allium hollandicum* 'Purple Sensation'.

Deciduous plant
A plant which sheds its foliage at the end of the growing season.

Evergreen plant
A plant that retains its foliage all year and sheds older leaves at intervals throughout the year.

F1 hybrid
A first generation offspring produced naturally or artificially from two genetically different plants. The 'F' stands for 'filial'. Although the resulting plants are often more vigorous and have more desirable qualities than either parent, the seed produced by an F1 hybrid is rarely of similar worth and seldom comes true to type.

Family
A grouping of plants classified according to the similarity of their flowers and fruit. A family can contain one or many (sometimes hundreds of) genera.

Frost hardy
A plant that can withstand temperatures down to −5°C.

Genus (genera)
A grouping of one to many closely related species, denoted by the first part of the Latin name, for example, *Allium.*

Half-hardy plant
A plant that cannot tolerate frost but can withstand temperatures down to 0°C.

Hardy plant
A plant that can withstand temperatures down to −15°C.

Herb
Generally defined as a plant that is valued for its flavour, fragrance, medicinal qualities, economic and industrial uses, pesticidal properties and colouring materials. In short, a plant which is deemed to be useful to mankind.

Herbaceous perennial
A plant with non-woody stems which dies down to nothing over winter but reappears each spring.

Hybrid plant
An offspring produced naturally or artificially from two genetically different plants.

Mulch
A layer of usually organic matter applied to the soil over or around a plant to conserve moisture, protect the roots from temperature fluctuations, reduce weed growth and enrich the soil.

Nectar
A sugar secretion produced from plant nectaries, which is attractive to pollinating insects.

Nectaries
The tissues or glands in plants that produce nectar.

Perennial
A plant that lives for more than two years and produces flowers each year.

pH
pH is an abbreviation of 'potential of Hydrogen'. This is a measurement which gauges how acidic or alkaline a material is. It is often applied to soil.

Pollen
The male reproductive cells of a plant.

Pollination
The transfer of pollen from the stamen to the stigma of the same or different flowers.

Rhizome
An underground horizontal stem that is usually swollen and fleshy.

Shrub
A woody perennial plant, usually with multiple stems.

Species (sp.)
The subdivisions of a genus, denoted by the second part of the Latin name. For example, *Prunus domestica*.

Stamen
The male reproducing organ of a flower, which contains the anther and filament.

Stigma
The upper part of the female reproductive organ of a flower, which receives the pollen.

Subspecies (subsp.)
Similar to variety. A subdivision within a species that has characteristics easily attributable to the species but differs in small but significant ways. For example, *Prunus domestica* subsp. *insititia*, the damson, is a subspecies of *Prunus domestica*, the plum.

Tender plant
A plant that is vulnerable to low temperatures.

Tree
A woody perennial plant with a single main stem or trunk.

Tuber
A swollen underground stem or root where food is stored.

Variety (var.)
Similar to subspecies. A subdivision within a species that differs consistently in small but significant ways. For example, *Rosa gallica* var. *officinalis*.

Index of Common Names of Plants and their Latin Equivalents

Plants whose common name is the same in Latin have not been listed.

Common Name	Latin Name	Page
African marigold	*Tagetes*	64
Almond	*Prunus amygdalus*	245
Alpine strawberry	*Fragaria vesca*	200–1
Alsike clover	*Trifolium hybridum*	130, 131
Apple	*Malus domestica*	71, 106, 138, 193, 201, 246
Apricot	*Prunus armeniaca*	245
Autumn crocus	*Colchicum autumnale*	112, 171, 242
Barberry	*Berberis*	177, 183, 244
Bay	*Laurus nobilis*	201, 209, 222, 248
Bee balm	*Monarda*	130
Bee-bee tree	*Tetradium daniellii*	107, 190, 191, 245
Bell heather	*Erica cinerea*	92, 93, 173
Betony	*Stachys officinalis*	159, 248, 249
Birdsfoot trefoil	*Lotus corniculatus*	249
Blackcurrant	*Ribes nigrum*	195, 201, 246
Blackberry	*Rubus fruticosus*	18, 20, 23, 199, 201, 246
Black-eyed Susan	*Rudbeckia*	30, 31, 80, 112, 168, 171, 172, 246, 288
Blackthorn	*Rhamnus catharticus*	245
Californian poppy	*Eschscholzia californica*	92, 142, 149, 241
Candytuft	*Iberis umbellate*	139, 142, 149, 228, 241
Catmint	*Nepeta*	4, 7, 92, 161, 243, 248
Cherry – sour	*Prunus cerasus*	245
Cherry – sweet	*Prunus avium*	71, 138, 194, 201, 245, 246
Cherry laurel	*Prunus laurocerasus*	176, 183, 244
Cherry pie plant	*Heliotropium*	140, 145, 146, 149, 236, 241

258

Chicory	*Chicorium intybus*	190, 248
China aster	*Callistephus chinensis*	96, 143, 149, 241
Chinese bee tree	*Tetradium daniellii*	107, 190, 191, 245
Chinese scholars' tree	*Sophora japonica*	107, 190, 191, 245
Chives	*Allium schoenoprasum*	98, 203, 209, 222, 248
Coltsfoot	*Tussilago farfara*	78, 210, 217, 239, 240, 249
Comfrey	*Symphytum officinale*	248
Coneflower	*Echinacea*	80, 171, 228
Coneflower	*Rudbeckia*	30, 31, 80, 112, 168, 171, 172, 228, 243
Cornflower	*Centaurea cyanus*	21, 35, 39, 96, 144, 149, 240, 241, 248, 249
Crab apple	*Malus sylvestris*	245
Damson	*Prunus domestica* subsp. *insititia* 71	
Dandelion	*Taraxacum officinale*	18, 21, 35, 40, 66, 78, 138, 210, 217, 238, 249
Deadnettle	*Lamium maculatum*	72, 153, 172, 235, 236
Devil's bit scabious	*Succisa pratensis*	35, 113, 249
Dill	*Anethum graveolens*	204, 248
Evening primrose	*Oenothera biennis*	248
False acacia	*Robinia pseudoacacia*	139, 245
Fennel	*Foeniculum vulgare*	203, 209, 248
Field poppy	*Papaver rhoeas*	19, 28, 96, 101, 154, 216, 249
Field scabious	*Knautia arvensis*	66, 92, 141, 217, 239, 240, 249
Fleabane	*Erigeron*	92, 142
Flowering currant	*Ribes sanguineum*	72, 116
Foxglove	*Digitalis*	130, 133
Globe thistle	*Echinops*	92, 94, 116, 164, 172, 228, 235, 236, 242
Goat willow	*Salix caprea*	21, 70, 185, 191, 245
Golden rod	*Solidago*	168, 243
Gooseberry	*Ribes uva-crispa* var. *reclinatum* 191, 201, 246	
Gorse	*Ulex europaeus*	40, 72, 117, 139, 173, 183, 244
Grape hyacinth	*Muscari*	70, 113, 152, 172, 243
Hawthorn	*Crataegus monogyna*	22, 72, 186, 191, 245
Heartsease	*Viola tricolour*	78
Heliotrope	*Heliotropium*	140, 145, 146, 149, 236, 241
Himalayan balsam	*Impatiens glandulifera*	30, 96, 249
Holly	*Ilex aquifolium*	116, 187, 191, 245
Horse chestnut	*Aesculus hippocastanum*	21, 72, 185, 191, 245
Horsetail	*Equisetum arvense*	49
Hyssop	*Hyssopus officinalis*	21, 66, 98, 208, 209, 222, 248

Ice plant	*Sedum spectabile*	80, 81, 164, 172, 228, 243
Indian bean tree	*Catalpa bignonioides*	86, 87, 188, 191, 245
Ivy	*Hedera helix*	109, 110, 244
Jacob's ladder	*Polemonium caeruleum*	72, 76, 159, 172, 228, 235, 236, 243
Japanese pagoda tree	*Sophora japonica*	107, 190, 191, 245
Joe Pye weed	*Eupatorium purpureum*	248
Jostaberry	*Ribes* x *culverwellii*	195
Judas tree	*Cercis siliquastrum*	245
Knapweed	*Centaurea scabiosa*	92, 113, 249
Lady's bedstraw	*Galium verum*	249
Lamb's ears	*Stachys byzantina*	92, 95, 159, 172, 243
Lavender	*Lavandula*	86, 88, 98, 100, 106, 138, 204, 209, 222, 235, 236, 244, 248
Leatherwood	*Eucryphia*	108, 109, 181, 183, 244
Lemon balm	*Melissa officinalis*	98, 205, 209, 221, 222, 248
Leopard's bane	*Doronicum*	72, 151, 152, 172, 235, 236, 242
Lesser celandine	*Ranunculus ficaria*	249
Lesser trefoil	*Trifolium dubium*	249
Lime	*Tilia*	26, 86, 189, 191, 245
Linden	*Tilia*	86, 189, 191, 245
Ling heather	*Calluna vulgaris*	92, 93, 108, 182, 183, 244
Livingstone daisy	*Mesembryanthemum criniflorum*	63
Loganberry	*Rubus* x *loganobaccus*	247
Loosestrife	*Lythrum salicaria*	249
Love-in-a-mist	*Nigella*	146, 149, 228, 236, 241
Mallow	*Malva*	249
Manuka	*Leptospermum scoparium*	90, 180, 183, 235, 236, 244
Marjoram	*Origanum vulgare*	22, 25, 66, 98, 138, 205, 209, 222, 235, 248
May	*Crataegus monogyna*	22, 72, 186, 191, 245
Meadow buttercup	*Ranunculus acris*	35, 249
Meadow clary	*Salvia pratensis*	66, 213, 217, 239, 249
Meadow cranesbill	*Geranium pratense*	216, 217, 239, 240, 249
Meadow geranium	*Geranium pratense*	216, 217, 239, 240, 249
Meadow sage	*Salvia pratensis*	66, 213, 217, 239, 249
Meadowsweet	*Filipendula ulmaria*	249
Medlar	*Mespilus germanica*	71, 196, 201, 246
Michaelmas daisies	*Aster* sp.	21, 80, 82, 112, 167, 172, 228, 235, 236, 242
Mint	*Mentha*	66, 98, 138, 206, 209, 222, 248
Mock orange	*Philadelphus*	180, 183, 244
Monkshood	*Aconitum*	130, 151

Mountain ash	*Sorbus aucuparia*	21, 72, 138, 187, 191, 245
Naked ladies	*Colchicum autumnale*	112, 171, 242
Nasturtium	*Tropaeolum majus*	248
Old man's beard	*Clematis vitalba*	249
Oregon grape	*Mahonia aquifolium*	70, 117, 175, 183, 244
Oriental poppy	*Papaver orientale*	19, 21, 92, 116, 154, 172, 235, 236, 243
Ornamental onion	*Allium*	242
Oxeye daisy	*Leucanthemum vulgare*	35, 37, 212, 239, 249
Peach	*Prunus persica*	138, 245
Pear	*Pyrus communis*	71, 106, 138, 194, 201, 246
Perennial daisy	*Leucanthemum*	92, 212, 243
Plum	*Prunus domestica*	71, 106, 138, 192, 201, 246
Poached egg plant	*Limnanthes douglasii*	147, 148, 149, 236, 241
Pot marigold	*Calendula officinalis*	96, 97, 248
Privet	*Ligustrum vulgare*	26
Quickthorn	*Crataegus monogyna*	22, 72, 186, 191, 245
Quince	*Cydonia oblonga*	71, 198, 201
Ragwort	*Senecio jacobaea*	26
Raspberry	*Rubus idaeus*	138, 198, 201, 247
Red clover	*Trifolium pratense*	32, 130, 131
Redcurrant	*Ribes rubrum*	195, 201, 247
Red deadnettle	*Lamium purpureum*	18, 21, 40, 78, 138, 210, 217, 239, 240, 249
Red hot poker	*Kniphofia*	165, 172, 235, 236, 243
Rocket	*Eruca vesicaria* var. *sativa*	98
Rose	*Rosa*	90, 91, 138, 179, 183, 244
Rosebay willowherb	*Chamerion angustifolium*	249
Rosemary	*Rosmarinus officinalis*	72, 138, 202, 209, 222, 244, 248
Rowan	*Sorbus aucuparia*	21, 72, 138, 187, 191, 245
Russian comfrey	*Symphytum* x *uplandicum*	47, 140
Russian sage	*Perovskia atriplicifolia*	86, 89, 108, 169, 172, 228, 235, 236, 243, 244
Sage	*Salvia officinalis*	66, 98, 138, 208, 209, 221, 248
Sainfoin	*Onobrychis viciifolia*	100, 139, 214, 217, 249
Sea holly	*Eryngium*	92, 94, 139, 160, 172, 228, 242
Sloe	*Prunus spinosa*	245
Snapdragon	*Antirrhinum majus*	22, 38, 41
Sneezeweed	*Helenium*	80, 81, 92, 162, 168, 172, 235, 236, 243
Snowdrop	*Galanthus nivalis*	70

Soapwort	*Saponaria officinalis*	248
Spanish chestnut	*Castanea sativa*	245
Starflower	*Borago officinalis*	21, 22, 24
Strawberry	*Fragaria* x *ananassa*	21, 200, 247
Summer savory	*Satureja hortensis*	98, 207, 248
Sunflower	*Helianthus annuus*	96, 98, 100, 138, 148, 149, 228, 241
Sweet chestnut	*Castanea sativa*	245
Sweet cicely	*Myrrhis odorata*	98, 248
Sweet clover	*Melilotus officinalis*	100, 103, 215, 217, 226, 239, 240, 249
Sweet mignonette	*Reseda odorata*	96, 147, 149, 215, 236, 241
Sweet rocket	*Hesperis matronalis*	248
Sweet sultan	*Amberboa moschata*	96, 149, 241
Sweet violet	*Viola odorata*	78
Sycamore	*Acer pseudoplatanus*	72, 245
Tansy	*Tanacetum vulgare*	248
Tea tree	*Leptospermum scoparium*	90, 180, 183, 235, 236, 244
Tea tree	*Melaleuca alternifolia*	90, 180
Teasel	*Dipsacus*	140, 141, 249
Thyme	*Thymus*	66, 207, 221, 222, 248
Traveller's joy	*Clematis vitalba*	249
Tulip tree	*Liriodendron tulipifera*	86, 245
Valerian	*Centranthus ruber*	112, 157, 242
Viper's bugloss	*Echium vulgare*	21, 140, 216, 217, 239, 240, 249
Wallflower	*Cheiranthus*	72, 77, 114, 139, 141, 149, 236, 241
White clover	*Trifolium repens*	22, 25, 32, 78, 100, 139, 213, 217, 239, 249
White willow	*Salix alba*	245
Whitebeam	*Sorbus aria*	138, 245
Wild carrot	*Daucus carota*	214, 217, 249
Wild mignonette	*Reseda lutea*	147, 215, 217, 239, 240, 249
Windflower	*Anemone blanda*	151, 172, 242
Winter aconite	*Eranthis hyemalis*	70, 114, 150, 172, 242
Winter savory	*Satureja montana*	98, 207, 209, 222, 248
Yarrow	*Achillea millefolium*	21, 92, 154, 155, 156, 249

Plant Index

Abelia sp. 244

Acer pseudoplatanus 72, 245

Achillea millefolium (page 155) 21, 92, 154, 155, 156, 249

Aconitum sp. 130, 151

Aesculus hippocastanum 21, 72, 185, 191, 245

Allium hollandicum (page 115) 114, 115, 158, 236

Allium schoenoprasum 98, 203, 209, 222, 248

Allium sp. 242

Alyssum sp. 241

Amberboa moschata 96, 149, 241

Anchusa azurea 158, 172, 242

Anchusa officinalis 140, 242

Anemone blanda 151, 172, 242

Anemone x *hybrida* 242

Anethum graveolens 204, 248

Angelica archangelica 248

Antirrhinum majus (page 41) 22, 38, 41

Arabis sp. 242

Arbutus unedo 244

Armeria maritima 242

Asparagus 247

Aster sp. (page 82) 21, 80, 82, 112, 167, 172, 228, 235, 236, 242

Aubretia sp. 242

Berberis sp. 177, 183, 244

Borago officinalis (page 24) 21, 22, 24, 98, 100, 140, 213, 217, 248, 249

Broad Bean 19, 139, 225, 247

Broccoli 223, 224, 247

Bryonia dioica 249

Buddleja davidii (page 85) 84, 85, 86, 116, 244

Buxus sempervirens 244

Cabbage 139, 223, 224, 225, 247

Calabrese 247

Calendula officinalis (page 97) 96, 97, 248

Callistephus chinensis 96, 143, 149, 241

Calluna vulgaris 92, 93, 108, 182, 183, 244

Caltha palustris 249

Carrot 65, 139, 214, 224, 226, 247

Caryopteris sp. 244

Catalpa bignonioides (page 87) 86, 87, 188, 191, 245

Cauliflower 139, 247

Ceanothus sp. 244

Centaurea cyanus (page 39) 21, 35, 39, 96, 144, 149, 240, 241, 248, 249

Centaurea scabiosa 92, 113, 249,

Centranthus ruber 112, 157, 242,

Cercis siliquastrum 245

Cerinthe major var. *purpurascens* (page 132) 130, 132

Chaenomeles x *superba* 244

Chaenomeles speciosa 72, 244

Chamerion angustifolium 249

Cheiranthus cheiri (page 77) 72, 77, 114,

139, 141, 149, 236, 241
Chicorium intybus 19, 248
Chicory 19, 247
Cirsium sp. 249
Clarkia sp. 143, 149, 228, 241
Clematis sp. 72, 178, 235, 236, 249
Colchicum autumnale (page 112) 112, 171, 242
Corylus avellana (page 184) 183, 184, 191, 245, 247
Cosmos bipinnatus (page 145) 118, 144, 145, 149, 228, 241
Cotoneaster sp. 86, 116, 138, 177, 178, 183, 244
Courgette (page 105) 54, 61, 102, 105, 224, 225, 247
Crataegus monogyna 22, 72, 186, 191, 245
Crocus sp. 70, 114, 150, 172, 235, 236, 242
Cucumber 247
Cucurbita pepo 30
Cydonia oblonga 71, 198, 201
Cynara cardunculus 242
Cytisus sp. 72, 244

Dahlia sp. (page 111) 80, 111, 112, 113, 118, 170, 172, 228, 242
Daphne mezereum 72, 175, 178. 183, 244
Daucus carota 214, 217, 249
Digitalis sp. (page 133) 130, 133
Dipsacus sp. (page 140) 140, 141, 249
Doronicum sp. 72, 151, 152, 172, 235, 236, 242
Echinacea purpurea (page 82) 80, 82, 116, 171, 172, 228, 235, 236, 242, 248
Echinops sp. (page 94) 92, 94, 116, 164, 172, 228, 235, 236, 242
Echium vulgare 21, 140, 216, 217, 239, 240, 249
Elaeagnus pungens 'Maculata' 51
Enkianthus campanulatus 244
Equisetum arvense 49
Eranthis hyemalis 70, 114, 150, 172, 242
Erica sp. 92, 93, 117, 173, 244
Erigeron sp. 92, 242
Eruca vesicaria var. *sativa* 98

Eryngium sp. (page 94) 92, 94, 139, 160, 172, 228, 242
Escallonia bifida 244
Eschscholzia californica 92, 142, 149, 241
Eucryphia sp. (page 109) 108, 109, 181, 183, 244
Euonymus japonicus 51
Eupatorium sp. 248, 249

Filipendula ulmaria 249
Florence Fennel 204, 224, 247
Foeniculum vulgare 203, 209, 248
Fragaria sp. (page 21) 20, 21, 138, 200, 201, 247
French Bean 247
Fuchsia sp. (page 108) 108, 244

Gaillardia sp. (page 163) 163, 242
Galanthus nivalis 70
Galium verum 249
Geranium pratense 216, 217, 239, 240, 249
Geranium sp. (page 76) 72, 76, 92, 156, 172, 235, 236, 243
Geum sp. 243
Godetia sp. 241

Hebe sp. 108, 244
Hedera helix (page 110) 109, 110, 244
Helenium sp. (page 81) 80, 81, 92, 162, 168, 172, 235, 236, 243
Helianthemum sp. 244
Helianthus annuus (page 98) 96, 98, 100, 138, 148, 149, 228, 241
Heliotropium sp. (page 146) 140, 145, 146, 149, 236, 241
Helleborus sp. 243
Hesperis matronalis 248
Hydrangea aspera 'Macrophylla' (page 9) 4, 9, 181, 183, 235, 236, 244
Hyssopus officinalis (page 67) 21, 66, 98, 208, 209, 222, 248

Iberis umbellata 139, 142, 149, 228, 241
Ilex aquifolium 116, 187, 191, 245
Impatiens glandulifera 30, 96, 249

Jasminum sp. 130
Juglans sp. 247

Kale 247
Knautia arvensis 66, 92, 141, 217, 239, 240, 249
Kniphofia sp. (page 166) 165, 172, 235, 236, 243

Lamium maculatum 72, 153, 172, 235, 236
Lamium purpureum 18, 21, 40, 78, 138, 210, 217, 239, 240, 249
Laurus nobilis 201, 209, 222, 248
Lavandula sp. (page 88) 86, 88, 98, 100, 106, 138, 204, 209, 222, 235, 236, 244, 248
Leek 224, 225, 247
Leptospermum scoparium 90, 180, 183, 235, 236, 244
Leucanthemum sp. 92, 212, 243
Leucanthemum vulgare (page 37) 35, 37, 212, 239, 249
Liatris spicata 21, 169, 172, 228, 243
Ligustrum sp. 26
Limnanthes douglasii (page 147) 147, 148, 149, 236, 241
Linaria vulgaris 249
Liriodendron tulipifera 86, 245
Lotus corniculatus 249
Lythrum salicaria 249

Mahonia aquifolium 70, 117, 175, 183, 244
Malus domestica 71, 106, 138, 193, 201, 246
Malus sylvestris 245
Malva sp. 249
Marrow 102, 247
Melaleuca alternifolia 90, 180
Melilotus officinalis (page 103) 100, 103, 215, 217, 226, 239, 240, 249
Melissa officinalis 98, 205, 209, 221, 222, 248
Mentha sp. (page 68) 66, 98, 138, 206, 209, 222, 248
Mesembryanthemum criniflorum 63
Mespilus germanica (page 197) 71, 196, 201, 246

Muscari sp. 70, 113, 152, 172, 243
Myosotis sp. 30, 140, 241, 243
Myrrhis odorata 98, 248

Nepeta sp. (page 7) 4, 7, 92, 161, 243, 248
Nigella sp. 146, 149, 228, 236, 241

Oenothera biennis 248
Olearia x haastii 84, 244
Onion 223, 224, 225, 246, 247
Onobrychis viciifolia 100, 139, 214, 217, 249
Origanum vulgare (page 25) 22, 25, 66, 98, 138, 205, 209, 222, 235, 248
Ornithogalum pyrenaicum 212, 239, 240, 249

Papaver orientale 19, 21, 92, 116, 154, 172, 235, 236, 243
Papaver rhoeas 19, 28, 96, 101, 154, 216, 249
Parsnip 139, 223, 224, 247
Penstemon sp. (page 36) 34, 35, 57, 116, 243
Perovskia atriplicifolia (page 89) 86, 89, 108, 169, 172, 228, 235, 236, 243, 244
Persicaria amplexicaulis 243
Phacelia tanacetifolia (page 101) 96, 101, 214, 217, 226, 239, 240, 249
Philadelphus sp. 180, 183, 244
Polemonium caeruleum (page 76) 72, 76, 159, 172, 228, 235, 236, 243
Potentilla fruticosa 244
Prunus amygdalus 245
Prunus armeniaca 245
Prunus avium 71, 138, 194, 201, 245, 246
Prunus cerasus 245
Prunus domestica 71, 106, 138, 192, 201, 246
Prunus domestica subsp. *inititia* 71
Prunus laurocerasus 'Otto Luyken' (page 176) 176, 183, 244
Prunus persica 138, 245
Prunus spinosa 245
Pulmonaria sp. 243

Pumpkin 102, 224, 247
Pyrus communis 71, 106, 138, 194, 201, 246

Radish 139, 246, 247
Ranunculus sp. 35, 249
Reseda lutea 147, 215, 217, 239, 240, 249
Reseda odorata 96, 147, 149, 215, 236, 241
Rhamnus catharticus 245
Rhododendron sp. 26, 52
Ribes nigrum 195, 201, 246
Ribes rubrum 195, 201, 247
Ribes sanguineum 72, 116
Ribes uva-cripa var. *reclinatum* 191, 201, 246
Robinia pseudoacacia 139, 245
Romneya coulteri 244
Rosa sp. (page 91) 90, 91, 138, 179, 183, 244
Rosmarinus officinalis 72, 138, 202, 209, 222, 244, 248
Rubus fruticosus (page 23) 18, 20, 23, 199, 201, 246
Rubus idaeus 138, 198, 201, 247
Rudbeckia sp. (page 31) 30, 31, 80, 112, 168, 171, 172, 228, 243
Runner Bean (page 104) 61, 102, 104, 139, 225, 247

Salix sp. 21, 70, 185, 191, 245
Salsify 226, 247
Salvia officinalis 66, 98, 138, 208, 209, 221, 248
Salvia pratensis 66, 213, 217, 239, 249
Salvia sp. 243
Saponaria officinalis 248
Sarcococca sp 244
Satureja hortensis 98, 207, 248
Satureja montana 98, 207, 209, 222, 248
Sedum spectabile (page 81) 80, 81, 164, 172, 228, 243
Senecio jacobaea 26
Skimmia japonica 116, 244
Solidago sp. 168, 243

Sophora japonica 107, 190, 191, 245
Sorbus aria 138, 245
Sorbus aucuparia 21, 72, 138, 187, 191, 245
Stachys byzantina (page 95) 92, 95, 159, 172, 243
Stachys officinalis 159, 248, 249
Succisa pratensis 35, 113, 249
Swede 224, 247
Symphoricarpos albus 86, 244
Symphytum sp. 47, 140, 248, 259

Tagetes sp. 64
Tanacetum sp. 248
Taraxacum officinale 18, 21, 35, 40, 66, 78, 138, 210, 217, 238, 249
Tetradium daniellii 107, 190, 191, 245
Thymus sp. 66, 207, 221, 222, 248
Tilia sp. 26, 86, 189, 191, 245
Trifolium dubium 249
Trifolium hybridum 130, 131
Trifolium pratense 32, 130, 131
Trifolium repens (page 25) 22, 25, 32, 78, 100, 139, 213, 217, 239, 249
Tropaeolum majus 248
Tulipa sp. (page 74–5) 72, 75, 113, 152, 236, 243
Tussilago farfara 78, 210, 217, 239, 240, 249

Ulex europaeus 40, 72, 117, 139, 173, 183, 244

Valeriana officinalis 249
Verbena bonariensis 161, 172, 228, 243
Verbena rigida (page 59) 57, 59, 116, 161, 172, 228, 236, 243
Viburnum tinus (page 73) 72, 73, 108, 174, 183, 235, 236, 244
Viburnum x *bodnantense* 72, 117, 174, 183
Viola odorata 78

Index

(Plants can be found in the separate
Plant Index)

annual plants 80, 92, 96, 102, 114, 228,
    236
    of value to bees 141–9, 241
aphids – *see* pests
Apis mellifera – *see* honey bee
Autumn 10, 40, 50, 54, 58, 62 105,
    106–16, 121, 142, 160, 171–2, 181,
    190, 208, 215, 238

bee
    bread 16, 20, 254
    brood xii, 10, 17, 20, 26, 27, 113,
        122, 124, 129 254
    garden 43, 44, 45
    hive (pages 231 and 233) xii, 2–3,
        4, 9, 10, 11, 13, 19, 22, 26, 27,
        28, 33, 43, 117, 129, 228, 230
biennial plants 72, 83, 141–9, 199, 214,
    215, 216, 236, 241, 255
bulbs 52, 55, 70, 232, 235–6, 255
    of value to bees 72, 150, 152, 157,
        212, 239, 242–3
    planting 113–14
bumblebee(s) xi, xiii, 1, 32, 35, 39, 83,
    96, 117, 121–5, 130, 134
    plants for 130

Chelsea chop 80
China 12, 190
climbing plants – *see* shrubs and
    climbers
Colony Collapse Disorder (CCD) 13
colour spectrum 28, 29, 40, 41
compost 46–8, 50, 57, 61
containers 52–3, 54, 114, 145, 193
corbicula 17, 18, 30, 254
corolla 22, 23, 32, 38, 39, 90, 130, 138,
    204, 255
cuttings 56-60
    basal 56, 57
    hardwood 56, 58, 116, 118
    root 56, 58-60, 116, 158
    semi-ripe 58, 105, 116
    softwood 56–7, 116

disease(s)
    bee 4, 13, 225
    plant 46, 114, 153, 223, 224

edible plants 102, 138, 139
    of value to bees 79, 191-201, 246–7

flowers – *see* individual entries in plant
    index
    double 38 64, 170

for cutting 142, 143, 148, 162, 168, 169, 223, 228

single 22, 38, 80, 90, 96

forage (foraging) 10, 18, 22, 26, 30, 40, 43, 64, 84, 107, 109, 122, 124, 151, 230, 232

fructose 20, 22

fruit 11, 13, 71, 79, 102, 116, 138, 191–201, 222, 223, 224–5, 246–7

gardening 43–66, 78-84, 102–6, 113–16, 117–19

organically 44–9, 83

Gerard, John 144, 152, 205, 208

glucose 16, 20, 22, 109

green manure 215, 226

ground preparation 49-50

herbicide 44, 49

herbs 43, 66, 96–8, 100, 105, 138, 139, 140, 225, 230

of value to bees 201–9, 248

planting plan 219–22

honey 3, 10, 15, 19, 22, 26, 27, 90, 92, 98, 202

gathering 2

crop 17, 22,

sac – *see* honey crop

stomach – *see* honey crop

honeybee(s)

development of 15, 16

drone 9, 10, 15, 16, 254

economic value of 12

farming 1, 2

feral 4

proboscis 17, 31–2, 38, 39, 254

pollen 'basket' – *see* corbicula

queen 2, 4, 9–10, 11, 15, 16, 254

types of 8

vision 28–30

worker 9, 10–11, 15, 16, 17, 26, 30, 33, 84, 255

honeydew 86, 189

hoverfly 121, 127, 128

hypopharyngeal gland 15, 16, 17, 19, 254

insecticide 12, 44

June gap 40

kitchen garden – *see* potager

labelling 62–3, 80

lawn 66, 126

manure 46, 47, 48, 50, 53, 61

green – *see* green manure

Markham, Gervase 43, 48, 141

mulch(ing) 46, 47, 55, 78, 113, 116, 162, 256

nectar 14, 15, 18, 20–6, 30, 32, 38, 39, 40, 49, 70, 72, 78, 79, 83, 86, 98, 100, 107, 109, 114, 121, 130, 138, 256

guide 34–5

poisonous 26

nectaries 15, 20, 22, 38, 256

organic gardening 44–9, 83

perennial plants 35, 57, 60, 72, 79–80, 92, 110, 112, 114, 228, 232, 235, 256

of value to bees 150–172, 242

pesticide 44, 83

pests 12, 45, 55, 86, 127, 129, 189, 223

planting plan(s) 219–40

3 seasons border 235–6

herb garden 220–1

potager 228–30

wild flower garden 239-240

plants

annuals – *see* annual plants

bare root 53, 118

biennials – *see* biennial plants

choosing 40, 193

deadheading 66, 105–6, 110, 156, 212

dividing 60

growing plants in containers 52–3, 54, 114, 145, 193

hardening off 62

herbs – *see* herbs

perennials – *see* perennial plants

wild flowers – *see* wild flowers

pollen 14-20, 26, 27, 30, 33, 38, 40, 70, 72, 83, 90, 100, 107, 113, 117, 121, 257

   basket – *see corbicula*

   colour 19, 21

pollination 11–12, 18, 30, 102, 257

   of commercial crops 11–13

potager (kitchen garden) 79, 222–30

   large, plan for 228

   small, plan for 230

Proboscis 17, 31–2, 38, 39, 86, 130, 254

protein 15–16, 129

royal jelly 15–16, 254

scent 30, 34

seed sowing 60–2, 83, 102, 114, 237

shrubs and climbers 51, 53, 56, 60, 64, 72, 79, 84, 86, 105, 106, 108–9, 116, 118, 232, 235, 257

   of value to bees 173–83, 244

skeps 2-5

soil 46, 47, 51–2, 53, 92, 220, 234, 237, 240

   pH value of 52, 256

solitary bee(s) xi, 1, 126–7, 130, 132

spring 14, 40, 54, 55, 56, 57, 58, 62, 69–84, 114, 122, 126, 141, 150, 152, 174, 175, 177, 183, 185, 187, 191, 196, 201, 203, 210

stamen 14, 15, 20, 38, 257

stigma 14, 15, 23, 257

sucrose 20, 22

summer 11, 22, 58, 62, 78, 84–106, 122, 142, 152, 157, 160, 177, 179, 181, 187, 188, 190, 196, 203, 208, 210, 213, 215

swarm 2, 3, 4-6, 10, 205, 255

temperature 10, 11, 14, 20, 26, 27, 51, 63, 65, 117, 151, 256, 257

tongue – *see* proboscis

tree(s) 4, 40, 53–4, 60, 72, 79, 86, 105, 107, 118, 257

   fruit 71, 106, 192–5, 196, 198, 224

   of value to bees 183–91, 245

ultraviolet 28, 29

urban heat island dffect 14

varroa mite 4, 255

vegetables xiii, 46, 54, 61, 139, 222, 223–4, 225, 246, 247

virgil 205, 219

waggle dance 33

wasp(s) 1, 45, 121, 127–9, 165

water(ing) xii, 14, 22, 26–8, 31, 48, 49, 52, 53, 54, 57, 58, 60, 61, 62, 80, 105, 238

   evaporation 26, 27

weed(s)/weeding 35, 46, 49, 50, 55, 102, 226, 237

wild flowers 34, 35, 66, 78, 83, 92, 113, 237–40

   of value to bees 210–17, 249

winter 9, 10, 11, 16, 22, 50, 54, 58, 90, 105, 107, 113, 116–19, 121, 126, 128, 174, 183